Niles Eldredge

AND THE

# THE TRIUMPH

FAILURE OF

# OF EVOLUTION

CREATIONISM

A Peter N. Nevraumont Book

W. H. FREEMAN AND COMPANY
New York

Created and Produced by
NEVRAUMONT PUBLISHING COMPANY, INC.
10 East 23rd Street
New York, New York 10010

A Peter N. Nevraumont Book

Cover and text design Tsang Seymour Design Inc., New York, New York

Library of Congress Cataloging-in-Publication Data
Eldredge, Niles.
      The triumph of evolution: and the failure of creationism / Niles Eldredge
        p. cm.
      Includes index.
      ISBN 0-7167-3638-1 (hardcover)
      ISBN 0-7167-4478-3 (paperback)
Human evolution. 2. Creationism. 1. Title.

GN281.4.E45 2000
599.93'8-dc21
           99-058515

Portions of this text have been previously published in the author's *The Monkey
Business: A Scientist Looks at Creationism* (Washington Square Press).

Printed in the United States of America

**First paperback printing 2001**

W. H. FREEMAN AND COMPANY
41 Madison Avenue
New York, New York 10010
Houndsmills, Basingstoke RG21 6XS, England

Dedicated to the inspired and unflagging efforts of Eugenie Scott and her entire staff at the National Center for Science Education—frontline defenders of quality science education in America.

# In the Beginning
## Religion, Science, . . . and Politics

That the United States and the rest of the modern world are fundamentally a secular, technologically based society (albeit one generally committed to the free and unfettered practice of religion) is nicely brought out by the Y2K doomsday myth so widely adopted as we approached the recent millennial date: January 1, 2000. The dark scenario of widespread shortages and other societal malfunctions born of computer glitches, after all, was universally seen as delivered not by a vengeful, wrathful God, but rather by us humans.

Early programmers had assumed (if they thought about it all) that their shorthand, two-digit system of keeping track of yearly dates would long since have changed by the year 2000. In sharp contrast, previous millennial myths saw doom and destruction as God's payback for our sins—still our fault, of course, but with punishment meted out by God, not by errant machines. Likewise, we thanked the techno-fixers—not a merciful God—that the worst of the Y2K problem was handily cut off at the pass. That we were able to blame computer programmers, and not God, for what seemed to so many as impending doom and still manage to concoct a millennial scenario of darkest catastrophe just as all our forebears crossing the previous millennial divide did, shows us how far we have—and haven't—come.

But if the doomsday scenario this time was completely secularized, nonetheless the advent of the Millennium has intensified contact between science and religion—much of it in the spirit of conciliation, though some of it with continued mistrust and hostility. Currently more than several hundred college courses specifically address "science and religion." The Templeton Foundation annually awards a sum in excess of that carried by a Nobel Prize in recognition of the furtherance of closer ties between science and religion. In 1999, for example, the award was $1.24 million compared to the more modest $978,000 handed out by the Nobel Committee in 1998 (though in fairness it must be said that there is only one Templeton Prize, whereas there are several Nobel Prizes). Numerous colloquia on science and religion have been held—some sponsored by religious institutions, such as the Vatican, and some by decidedly secular institutions, such as the American Association for the Advancement of Science. Television shows and, of course, many books and articles have been in full cry as well.

I see several distinct ways in which science and religion are variously engaged either in potentially fruitful dialogue, or at daggers drawn or simply as ships passing in the night. The latter relationship is simply stated: in most nations—technologically advanced or impoverished, agrarian Third World alike—there is little day-to-day contact between the realms of science and religion. That is as true of the United States as it is of most of the nations of Europe, South America, Asia, and Africa. In countries where forms of Christianity predominate, for example, the overwhelming mainstream has, for well over a century, viewed the relation of science and religion as essentially neutral: each constitutes an important sector of society, but each does a vastly different job.

From this perspective, the role of religion is spiritual, moral, and social. Science, on the other hand, is there to discover the workings of the universe—and to lead to technological advance. This is why so many scientists (such as my friend and colleague, paleontologist Stephen Jay Gould) advocate a polite going of separate ways—a sort of benign acknowledgment that each exists, but can and should have little to do with one another. That is the general stance that I myself have adopted in my earlier works on creationism in American society—a sort of "rendering unto Caesar" division of labor that would minimize conflict but at the same time not look for any particular close resonance between the two domains.

But others insist that there is either resonance—or inherent conflict—between the domains of science and religion. I believe my colleague Margaret Wertheim[1] is right when she says that, in Western culture, historically speaking, the supposed warfare between science and religion has been greatly exaggerated. Indeed, most of the formative figures in the emergence of modern science were deeply religious and thought (as Wertheim has observed) that they were discovering the "mind of God" every bit as much as some modern physicists appear to think they are. Yet it is undoubtedly true that with the Darwinian revolution of the mid-nineteenth century—with the certain knowledge gained by some newer branches of science that the Earth is very old, has had a very long history (and especially that life on Earth is almost as old), and has had a history of change—has collided very deeply with conservatively held traditional religious beliefs in Judeo-Christian circles.

There are, indeed, many people who believe literally that the notion of biological evolution is the work of the devil.[2] As detailed here, I have spent over twenty years talking with, debating, and reading the literature of creationists—roughly, people who believe that God created the heavens and Earth, and all living things according to accounts in Genesis. I remain convinced that their unrelenting hatred of the very idea of evolution stems from their concept of morality: where morals come from, and why people behave in a moral fashion (when they do). The argument is simple: the Bible says that "mankind" was created in God's image. If that is not true, if instead we are descended from the apes, then there is no reason whatsoever to expect humans to behave in a godlike, moral fashion. We would, instead, be expected to behave like "animals." The conviction is deeply held.

Thus, in some quarters, it is simply not possible to assign to science the task of cosmology while giving religion the role of articulating a moral and spiritual understanding of what it means to be a living human. It is not possible because the two are seen as inseparable: morality flows automatically and solely from the manner in which humans were "created" in the first place. From this perspective, religion (meaning, specifically, certain forms of religion—especially conservative Christianity, but also conservative strains of Islam and Judaism[3]) are fundamentally at odds with at least some forms of the scientific enterprise.

In the United States especially, creationism is associated not only most closely with aspects of Christian Fundamentalism, but with conservative (mostly, if not exclusively, conservative Republican) politics. And though I document this charge fully in later chapters of this narrative, I cannot emphasize enough at the outset that politics is the very essence of this conflict. It is the belief that evolution is inherently evil—a belief that stems from religious interpretation, *and therefore poses a threat to the hearts and minds of the populace,* that, I am convinced, motivates the vast majority of the creationists. Thus the issue is about what is to be taught in the public schools, and the arena in which the battle takes place goes far beyond local school board meetings and classroom confrontations: it includes bills passed by state legislatures and opinions handed down by the Supreme Court of the United States. It includes judgments passed by statewide school boards, such as the decision to downgrade the teaching of evolution in the statewide syllabus in Kansas late in the summer of 1999—a mandate issued after the first text of this narrative had already been written. On the face of it, then, creationism is a political issue—and has been at least since Clarence Darrow defended John Scopes against the prosecutorial zeal of William Jennings Bryan in Dayton, Tennessee, on July 10–21, 1925.

Are there creationists who are religiously motivated but are not at the same time social and political conservatives? There must be, but in twenty years I have yet to encounter a single such person. Are there creationists, politically conservative or not, whose main concern does lie in the apparent moral implications of evolution and what it means especially to their own personal lives—whose main goal is not to influence what other people's kids are taught in school? Again, probably so. But the vast majority of active creationists do not restrict their activities to preaching to the converted, though they do plenty of that as well. They are motivated primarily to see that evolution is not taught in the public schools of the United States.

In any case, what creationism is *not* is a valid intellectual argument between opposing points of view. That battle was fought—with evolution emerging triumphant—in the latter half of the nineteenth century. Some twenty years ago, it was fashionable for creationists to claim that their views are not a religious, but rather a legitimate *scientific,* body of "knowledge." I will be taking a long hard look at the central claims of this so-called scientific creationism—the wolf in sheep's clothing concocted in the 1970s that deliberately removed religious rhetoric from the cant of creationism—a move calculated to bypass any objections based on the First Amendment to teach such patently religiously inspired material in a public school science class. No one was fooled. The best comment I have ever heard anyone make about scientific creationism came from Judge William Overton, who presided over the famed Arkansas trial in 1981: If this stuff is science, why do we need a law to teach it? The law in question was Arkansas's then recently passed "equal time" bill mandating that two competing versions of science ("evolution science" and "creation science") were equally valid, both deserving time and attention in the curriculum. Framed as an intellectually valid debate between supposedly opposing legitimate sets of scientific claims, instead scientific creationism was a shallow ploy. Intellectually, the debate has been dead since 1859—and evolution was triumphant!

Yet the debate rages on, and though tactics have changed, and once again creationists have become more open in acknowledging their religious motivations (often preferring now to claim that the idea of biological evolution is really a form of religious belief, rather than scientific theory), nonetheless the same old arguments *against* the validity of evolution as a scientific concept are still being trotted out. And thus, once more unto the breach, it becomes necessary to defend the integrity of evolution as a well-established body of knowledge and theory in science—intellectually triumphant not only within biology, or science generally, but within the intellectual framework of Western culture generally.

The newest, and by far most successful, voice in the creationist firmament belongs to Phillip Johnson, Boalt Professor of Law at the University of California, Berkeley. As we shall see in some detail (especially in Chapters 5 and 6), Johnson claims to have introduced something new into the debate: his conclusion that science in general, and evolutionary biology perhaps in particular, is rooted in what he calls philosophical naturalism, meaning that scientists think that the material world—matter and energy—is all that exists, and that an explanation for all natural phenomena necessarily entails only, well, natural phenomena. In other words, Johnson says that science is intrinsically and inherently atheistic. What, he says, if God really does exist, moreover the kind of proactive God in whom Johnson himself professes belief—one who is involved with details of everyone's daily life? Shouldn't the scientific enterprise be concerned that, given that possibility, the explanation of natural phenomena in terms of cause and effect that is the daily stuff of science might be hopelessly misguided if a divine agency outside the system were really pulling all the strings? Johnson finds it absurd that science could afford to ignore so cavalierly what might possibly be the real mover and shaker behind absolutely everything that happens.

Johnson turns a deaf ear to the obvious rejoinder: science is a human enterprise devised to experience in systematic ways the material universe. Everyone (including Johnson) agrees that a physical universe exists (actually, some people, though none of them creationists, have expressed doubt over even this proposition). We humans can directly experience that material world only through our senses, and *there is no way we can directly experience the supernatural.* Thus, in the enterprise that is science, it isn't an ontological claim that a God such as Johnson envisages does not exist, but rather an epistemological recognition that even if such a God did exist, there would be no way to experience that God given the impressive, but still limited, means afforded by science. And that is true by definition.

Johnson has a string of admirers drawn from academe, including philosophers and various scientists—among the latter, biochemist Michael Behe. I have met several of them and have "debated" both Johnson and Behe on college campuses (albeit only one time each). As I will recount in later chapters, beyond Johnson's charge of "philosophical naturalism" there is literally nothing new in their antievolutionary rhetoric. Their central thesis—that there are phenomena in the natural world of such great complexity that they simply cannot be explained by recourse to known natural processes—is exactly the same argument that was thrown at Darwin by the cleric St. George Jackson Mivart (1827–1900), one of the first and most ferocious critics of Darwin's *On the Origin of Species.* Mivart asserted that the human eye is simply too complex a structure to have evolved gradually and piecemeal, and therefore must instead be the intricate handiwork of a Creator.

Johnson, along with conservative members of Congress (such as House Republican Whip Tom DeLay) assert we are in the midst of a "culture war." Given the ethnic and cultural diversity that is the United States, given the highly secularized flavor of our big and blowsy society as epitomized by our attribution of Y2K doom at the Millennium to humans and *not to a vengeful God,* this proclaimed culture war is unilaterally declared by some segments of conservative Christian society in the population. But they are powerful, and they are continuing to have major impact. And that is why I feel I must once again pick up the cudgels to rail against these creationists—odd as it may seem, given that they have nothing new to say, and given that all the important issues as *intellectual issues per se* were resolved fully in the nineteenth century.

I take up this task because I am convinced that the integrity of science education in the United States and abroad is directly threatened by such nonsense. The issue is *not* whether any one particular student *believes* in evolution. But, as I develop throughout this book, it very much does matter that soon-to-be-adults living (and voting) in a technologically driven world know something about the ways their fellow humans gather knowledge about the natural world. Kids, in other words, need to be taught what science is all about—and that includes being exposed to the grandest conclusions of science. Not only do we need to continue to produce homegrown scientists, but also we need in this secular age to produce new generations of citizens who are conversant with science enough to make intelligent decisions in the polling place. Pretending to young minds that we cannot tell the difference between good science and bad, between the real and the bogus, not only sends a horribly distorted message about the very nature of science, but also makes evident to most students that adults don't care much about the truth. I write this book because those who see a necessary conflict between science and religion—and a "culture war" over the hearts and minds of the American populace—are doing their best to destroy quality science teaching in the United States.

There are still other ways in which science and religion commingle in modern American life—ways that see at least potential common ground, rather than inherent conflict, or simply benign nonintersection. In a major aspect of this common-ground approach, we are treated to visions of physicists contemplating the mind of God, and some theologians looking for God in scientific data. I once read an account of a meeting in which the consensus was reported to be that if the rate of expansion of the universe is greater than the escape velocity of stars fleeing the center of the universe from the Big Bang, then the universe is a singularity, literally a once-in-all-time phenomenon—as such casting doubt on the existence of God; if, instead, expansion rate is lower than escape velocity, expansion will slow down and the universe will eventually collapse back in on itself and, presumably, burst forth again, perhaps in an endless series of big bangs and implosions. This scenario suggested to the assembled scientists and theologians that perhaps a supernatural being lay behind the universe after all. When meant, as most of them are, in the spirit of ecumenicism, such exercises can do little harm, but I seriously wonder if they will ever shed much light on the nature either of God or of the universe.

Then there is the fact that many Western scientists, despite the common preconception, are themselves religious. In addition, many ordained clergy are also Ph.D.-holding, practicing scientists. Together these facts show that, to many, there is far from an intrinsic barricade between science and religion, though most religious scientists, no doubt, compartmentalize their workaday experiences from their religious beliefs and practices. Once again, whatever the nature of positive interactions in the personal lives of scientists or of the content of their professions and their religious beliefs, I see as yet no sign of any lasting benefit to either science or religion—except for the happy knowledge that the two can coexist within a single human breast.

I have held out what I take to be the best for last. For though I write this book determined to fight off the grimmer, darker messages attacking science in the name of religion, I really do see an as yet undervalued, as yet largely unexamined arena where lies, I think, some real hope for resonance between science and religion—and by "religion" I mean absolutely all religions that have ever existed of which we have any knowledge at all.

Creationism is just one of the two important social issues that have intersected my life as an evolutionary biologist. The other one is the horrendous loss of species—an event now gripping the planet that some have called the Sixth Extinction. Human beings are laying waste the world's ecosystems, and in so doing driving something like thirty thousand species of microbes, fungi, plants, and animals extinct every year. That's by far the fastest rate of ecosystem destruction and species loss since the time when the dinosaurs and so many other kinds of life were abruptly erased 65 million years ago—the result of a collision between Earth and extraterrestrial objects, and huge volcanic eruptions as well.

I have explored the question, How have humans entered the extinction business? In a series of three books, several articles, and a major exhibition at the American Museum of Natural History. I have proposed that culture became more important than traditional biological adaptations in the way ancestral humans approached the general problem of making a living. But the real change came when humans invented agriculture—and instantly became the first species in the entire 3.8-billion-year history of life to stop living inside local ecosystems. Among many other consequences, this move, clever as it was, took the normal controls off our population size, and we have skyrocketed from some 6 million people ten thousand years ago at the dawn of agriculture, to over 6 billion now. And that is the problem: 6 billion people, vying for unequally distributed wealth, are wreaking ecological havoc all over Earth.

And I have learned something more: people's concept of the gods or God change over time as well. And when examined in detail, concepts of God seem to dovetail remarkably well with my Western analysis of the relation of a given people to the natural world. I have come to see that religious traditions in general—and concepts of God in particular—reveal a lot about how people see themselves, and how they see themselves fitting into the natural world. It turns out that, naturally enough, people do tend to have a pretty clear idea of who they are and how they fit into the natural world in a functional sense (it's just their stories of how they got that way that tend to be fanciful!).

Thus a history of concepts of God should yield a pretty interesting human ecological history. And it suggests something more: if it is indeed the case, as I firmly believe it is, that this mounting loss of species and the accompanying topsoil loss and lack of adequate supplies of fresh water constitute some of the direst threats facing humanity right now, practitioners of the world's religions, many of whom are already aware of the environmental threats to their own lands, can potentially stand as the greatest source of good for the planet. Here, then, is a true millennial issue: a set of environmental problems besetting humanity at the year 2000, but a problem in which science and religion, instead of acting as enemies, stand a good chance of working together within the larger body politic to effect some truly positive measures. And though I plan to explore this positive side of the interaction between science and religion in book length form elsewhere, I cannot resist ending this present anticreationist tract with a preliminary exploration of these themes, which I do in the final chapter.

So far, I've talked about other people's views, and though I've made it clear that I am an evolutionary biologist not about to buy the possibility that people didn't evolve but rather were created by God sometime within the last ten thousand years, I have not revealed my own religious position. Here it is: I am a "lapsed Baptist." Along with many others, I see myself as an agnostic because "atheist" is too definitive, implying one can know something that is in principle unknowable. I will say that I am extremely skeptical that the kind of all-knowing, all-caring, all-doing God pictured in some circles exists. On the other hand, I think that concepts of God—*all* concepts of God—are about something, and of course I am not about to quarrel with anyone's personal interpretation of any one of those particular concepts of God. At the end of Chapter 7, I'll have more to say on this.

I confess also that I have personal reasons to become involved in the fight against creationism. One day in August of 1979, I received a phone call from a member of the Iowa Education Department. He wanted to know if I really had said that I thought that it would be a good idea to teach creationism alongside of evolution in high school classrooms.

I was appalled. I told him I had never said any such thing, and he replied that, in a typescript of an interview conducted by one Luther Sunderland with me in my office at the American Museum of Natural History, I was quoted as saying precisely that. Sunderland, an engineer by trade, had come to my office, representing himself as a "consultant" (how official that title was I never learned) to the New York State Board of Regents as they were conducting a curriculum review. He was really a lobbyist for creationism—and by all odds the most clever and successful creationist spokesman in the eastern United States in the late 1970s and early 1980s. Sunderland had sent me the typescript of our recorded conversation—with the invitation to change anything I felt was inaccurate or simply didn't like. The problem was, he left immediately for Iowa and introduced the uncorrected transcript into a legislative study session before I had received my copy and had a chance to edit it. The damage was done, and I felt humiliated as I listened to the Iowa official gently tell me that, however earnestly honest scientists try to be, they have to be aware of the political fallout of their remarks.

I was furious, of course, for letting Sunderland dupe me.[4] But it was a valuable learning experience: I learned right then that *the entire issue of creationism* (then largely masquerading as scientific creationism) *is purely a political battle*—for the hearts and minds of the nation's youth. I also learned that all was considered fair in the creationist wars. Determined to even the score, I wrote a piece for the journal *The New Republic* ("Creationism Isn't Science") in 1981—a brief salvo that caught the eye of a prescient young editor who invited me to expand it into book format. The result was *The Monkey Business* (1982), the forerunner to this book, written to set the record straight, and to provide ammunition to the hands of the nation's beleaguered high school biology teachers and their students—and for anyone else who might want a road map to help them navigate through creationist rhetoric—and what to say in reply.

In short, scientists cannot afford to shy away when broad social issues intersect their professional lives. I truly wish this updated and expanded version of my earlier consideration of creationism were not necessary. We simply have other, better things to do: more interesting things to think about in evolutionary biology and ecology, as well as pressing problems brought on by the calamitous decline of ecosystems and loss of species all around us to solve. But creationism *is* here, so we must fight! Evolution is triumphant in the intellectual realm, but it is still under siege in the political arena of the United States at the Millennium.

# CHAPTER 2

# Telling the Difference
## Science as a Way of Knowing

A favorite ploy of creationists as they argue to inject their particular brand of religion into the classroom is that, when you get right down to it, science and religion are alternative belief systems, and to be fair we should just get up in front of the kids, dispassionately recount both versions, and let the kids decide which version of events appeals to them more. They want us to tell our children that there is no way to distinguish between accounts of the origin and history of the universe, of the Earth, and of life derived from religious tradition as recounted in the Bible, on the one hand, and the latest thinking on these subjects in science, on the other. Insisting that there is no intrinsic difference between scientific and other forms of explanation—calling both, instead, alternative belief systems—cuts to the very heart of the creationist threat to educational integrity: if we tell kids there is no difference, they don't stand much of a chance of grasping the rudiments of science and how it is done.

Testability lies at the very heart of the scientific enterprise: if we make a statement about how things are—for instance, how the Earth originated, and how all the world's living species have come into being—we must be able to make *predictions* about what we logically should observe in the material world if that statement is true. Repeated failure to confirm predicted observations means we have to abandon an idea—no matter how fondly we cherish it, or how earnestly we may wish to believe it is true.

The history of science is littered with discarded ideas—notions of how things are that were simply not borne out by continued investigation. Science is far from being a belief system, for in science no idea is sacred. No statement is the ultimate truth. It is in the very nature of things that precious few ideas put forth to date in science have entirely withstood the test of time. Biological evolution is one of those ideas. So is the idea that the Earth is round.

## Facts and Theories: Is the Earth Really Round?

When Ronald Reagan injected creationism into the 1980 presidential campaign, he took a familiar route. Referring to evolution, he told reporters (after speaking to a group of fundamentalist clergy in Dallas, Texas), "Well, it is a theory, a scientific theory only, and it has in recent years been challenged in the world of science and is not yet believed in the scientific community to be as infallible as it once was believed."[1]

Beyond the fascinating commingling of politics, religion, and science, Reagan's remark picked up a standard creationist ploy when he said that evolution "is a theory, a scientific theory only." It is true that most of us in normal, daily life use the word "theory" to mean a tentative, sketchy notion about why or how something happened. All of us, for instance, have our own "theories" on why a band of southern Republicans in the House of Representatives fought so hard to oust Bill Clinton as President of the United States, or why the United States has been fighting in the Balkans— or why so many suburbanites love to drive SUVs. This is standard usage.

But creationists, including some who claim bona fide scientific credentials, have exploited the vernacular connotation of the word "theory," in effect saying that scientists use this word in precisely the same colloquial way. Thus, if evolution is "only a theory," our confidence in it ought to be less than if it were, say, a fact. Theories turn into "harebrained ideas" with ease. "Theory" is a bad word: to call an idea a theory is to impugn its credibility.[2]

Yes, theories in science are ideas: a theory may be a single, simple idea or, more usually, a complex set of ideas, as, for example, the Alvarez hypothesis invoking extraterrestrial impacts to explain the mass extinction of dinosaurs and a myridad of other organisms event at the end of the Cretaceous Period some 65 million years ago. There is no hard-and-fast way of telling a hypothesis apart from a theory; like lakes and ponds, and rivers and streams, theories and hypotheses differ in some vague way only by their size. Theories are generally grander, more encompassing than more narrowly focused hypotheses. But I have seen some mighty big ponds and some rather small lakes in my time.

Some theories are good and have withstood the test of time well. The idea that all organisms, past and present, are interrelated by a process of ancestry and descent—evolution—is such a theory. On the other hand, some theories have stood the test of time poorly and are no longer credited with much explanatory power. Spontaneous generation—the idea that organisms sprang from inorganic beginnings de novo, and are not all interrelated—has long been discarded as a useful scientific notion. It is taught in schools today, if at all, only as a historical curiosity.

Philosophers of science have argued long and hard over the differences among facts, hypotheses, and theories. But the real point is this: they are all essentially the same sort of thing. All of them—be they facts, hypotheses, or full-blown theories—are *ideas*. Some ideas are more credible than others. If the overwhelming evidence of our senses suggests that an idea is correct, we call it a fact. But the fact remains that a fact is an idea.

Let's take a concrete, if extreme, example that brings home this point very clearly. Consider the statement "the Earth is round." Is it a fact, a hypothesis, or a theory? A prominent creationist with whom I once spoke took offense at my suggestion that dismissing evolution as a credible notion was no different in principle from denying that the Earth is round. To him, and to most of us, that the Earth is round is a fact—something we all know to be true, something we take for granted. But why do we all think that "the Earth is round" is a fact? How many of us can perform a critical experiment to show that the Earth really is round? How many of us have ventured high enough into the upper reaches of the Earth's atmosphere that we could really see the Earth's curvature? Most of us have seen photos of the Earth taken from satellites, from spaceships, and from the moon. Clearly the Earth is round. But the relatively few vocal "flat-Earthers" have a counter even for this: to them, the spectacular achievements in space of the past half century are all an elaborate hoax—nothing more. To them, all the photos of the "Big Blue Marble" taken from satellites, space stations, and the moon itself are fakes.

Now, if the Earth is round, it is probably safe to assume it has always been so—at least since the dawn of human history, when we can pick up a written record of humanity's views on the question.[3] Yet the roundness of Earth was certainly not generally accepted as fact when Columbus set sail with his fleet of three ships. Indeed, many people thought it was a harebrained idea, and that Columbus was about to sail over the edge of what was patently a flat Earth. Only after the globe had been safely circumnavigated a number of times without a single ship dropping off the world's side did the roundness of the Earth start to take on the dimensions of credibility we deem necessary for a notion to become a fact.

Yet Eratosthenes, a Greek living in Ptolemaic Egypt in the third century b.c., showed that the Earth could not be flat with a simple yet conclusive experiment. His predecessors had already suggested the Earth is round because it casts a curved shadow on the moon. And ships sailing toward an observer appear on the horizon from the top of the mast down, also suggesting that the Earth is curved. Hearing that the sun shines directly down a well at Syene (now Aswan, Egypt) at noon on the summer solstice (the longest day of the year), Eratosthenes measured the angle between the sun's rays and a plumb bob he lowered down a well in Alexandria, some 600 miles north of Aswan, precisely at noon. That there was an angle at all in Alexandria was inconsistent with the idea that the Earth is flat. Eratosthenes could explain the phenomenon only if he envisaged a ball-shaped Earth. Using simple trigonometry, he calculated the circumference of the Earth to be the equivalent of about 28,000 miles, a respectable approximation to the 24,857 miles our modern instruments give us today. Columbus was aware of this and of later calculations, and he used them in his navigation.

Is the proposition that the Earth is round a fact, a hypothesis, a theory, or a downright falsehood? Obviously, it is an idea that has been variously considered all four. It was first called a wild idea, then a necessarily true conclusion (albeit accepted by only a few Greek savants); its respectability as a credible idea grew with the Renaissance. Now most of us proclaim it as fact—inasmuch as all attempts to disprove it have utterly failed. Flat-Earthers notwithstanding, we now even have direct confirmatory photographic evidence that the Earth is a sphere. But a round Earth is still an idea, albeit an extraordinarily powerful idea.

So what of Ronald Reagan's remark that evolution is "only a theory"? The answer is this: all of science is only a theory. And to label an idea as a theory in science is really a compliment, not a pejorative: for an idea to be called a theory in science, it has to have already passed many hurdles—and to look like it has a really good shot at being right.

## Evolution: How Good an Idea Is It?

The common expression "evolutionary theory" actually refers to two rather different sets of ideas: (1) the notion that absolutely all organisms living on the face of the Earth right now are descended from a single common ancestor, and (2) ideas of *how* the evolutionary process works—how, for example, do new species arise from old ones, and what processes actually underlay the reduction from four toes to but a single digit on the front feet of horses during the course of their 50-million-year evolutionary history? When scientists think about evolution in the first sense—i.e., has it actually happened—they strongly agree that it has, and many pronounce evolution in this first sense to be a fact. On the other hand, though biologists are in agreement on many of the basic mechanisms of the evolutionary process (the second sense of the expression "the theory of evolution"), many of the details are still being debated, as is healthy and normal in the unending human endeavor that is science.

Creationists love to gloss over this rather clear-cut, simple distinction between the idea that (1) life has evolved, and the sets of ideas on (2) how the evolutionary process actually works. Indeed, they like to pretend that disagreements and debates among biologists on the mechanisms of evolution somehow cast doubt on the first proposition—the fundamental idea that all life has evolved, that all species are descended from a single common ancestor in the remote geological past. That is exactly the import of Ronald Reagan's remark quoted earlier.

I will keep these two separate senses of evolution entirely distinct. In the remainder of this chapter and in the next, I will examine two grand predictions about what we should observe in the natural world if it is true that life has evolved. After demonstrating that the simple assertion that all species on Earth are interrelated passes all tests to falsify it with flying colors—and therefore that the theory of evolution in the first sense is as highly corroborated as any notion in science possibly can be—I will turn in Chapter 4 to a consideration of evolution in the second sense: what science has to say about *how* life evolves.

## The First Grand Prediction: Evolution Did Happen

Creationists are fond of pointing to the obvious fact that events that happened in the past are not subject to experimental verification or falsification, or to direct observation. After all, goes the creationist cry, no one was there at the beginning of the Cambrian Period to witness firsthand the supposed initial burst of evolutionary activity leading to the rapid evolution of complex animal life. How can we study something scientifically that has already happened? Creationists also note that few reputable biologists seem willing to predict what will happen next in evolution. And after all, says the creationist, if evolution is truly a scientific theory, it must be predictive—in the narrow sense of "making statements about what the future will hold" (and, of course, inherently untestable to biologists living in the moment). According to this creationist interpretation of science, that biologists neither can nor will predict the evolutionary future is strong evidence that the very idea of evolution isn't really scientific at all.

All this fancy rhetoric beclouds the simple meaning of "predictivity" in science. All that "predictivity" really means is that if an idea is true, there should be certain consequences—certain phenomena that we would expect—*predict*—to find if we looked. We should be able to go to nature—to the physical, material world—to see whether or not these predicted phenomena are really there.

So, in this spirit, we simply ask, If the basic idea is correct that all organisms past and present are interrelated by a process of ancestry and descent that we call evolution, what should we expect to find in the real world as a consequence? These observable consequences are the predictions we should be making—not guesses about the future.

Prediction I: The very idea of evolution—descent with modification—implies that some species are more closely related to each other than they are to more distant relatives. Therefore, we would predict that the living world is organized into groupings of closely similar species that are in turn parts of larger groups of more distant relatives that share fewer similarities, that are in turn parts of still larger groups with definite, if fewer, similarities. Eventually, the largest grouping of all—*all of life*—should be united by the shared possession of one or more characteristics. In other words, if evolution is true, the living world should be organized in a hierarchical fashion of groups within groups—a direct reflection of how closely related to one another each organism is.

In a very real sense, this prediction was discovered to hold true long before the idea of evolution was commonly accepted as the explanation for how the living world is organized. For at least a century before Charles Darwin (1809–1882) published *On the Origin of Species by Means of Natural Selection*, in 1859, biologists had recognized that life is organized into distinct groupings arranged in a natural, hierarchical fashion. The famous Swedish naturalist Carl von Linné (1707–1778)—more familiarly known simply as Linnaeus—had published the tenth edition of his *Systema Naturae* a full 101 years earlier, in which he outlined his scheme of classification of living things. Biologists today are still using the Linnaean hierarchy, which Linnaeus established as *a natural system, before the idea of evolution had been generally accepted*. (Linnaeus, like most other biologists before Darwin, was himself a creationist.) Linnaeus saw natural groupings of different kinds of plants within his Kingdom Plantae, and of different kinds of animals within his Kingdom Animalia. Biologists since Linnaeus's time have greatly refined his work, cataloguing hundreds of thousands of additional species and adding to the categories of Linnaeus's original classification scheme, but the basic hierarchical structure of Linnaeus's scheme remains—as it is simply a reflection of how biological nature is organized.

Darwin came along and simply showed why the Linnaean hierarchy exists—why it must be there if life has evolved. The Linnaean hierarchy, even though its rudiments were recognized almost a century before Darwin's epochal book, is a necessary consequence—a *prediction*—of what the structure of the living world must look like if all of life has descended from a single common ancestor.

Let's look at this prediction from a different perspective—literally, from the bottom up: Because there are a lot of *differences* between, say, bacteria, pine trees, rats, and humans, if evolution has actually happened, it must be the case that as new species arose from old, changes in the genetic, anatomical, and behavioral properties of organisms appeared from time to time. Later descendants would inherit these changes, while ancestors (whether survivors to the present, or found as fossils) would, of course, lack these new features. Because there are millions of species on the planet, we know that if evolution has occurred, there must be a process of lineage splitting—diversification—going on. The more recent the evolutionary diversification, the more similarities ought to be shared by organisms.

Here, a simple analogy drawn from human affairs is illuminating. Consider the work of patient monks in the Middle Ages who laboriously copied manuscripts from remote times and thus saved us from knowing even less than we do about our ancient past. From time to time a monk would make a minor mistake as he copied—a happy circumstance, it turns out, for historians whose job it is to track down the development of modern versions of ancient texts, for each undetected mistake was faithfully copied by later generations. Here we have descent: manuscripts being copied, and the copies being copied later. We also have modification: an early manuscript, free of errors, resembles its descendants to varying degrees. *An error introduced into a copy is passed on to all subsequent copies.* The result: the subsequent copies of manuscripts share more novelties (newly introduced items of change) than their earlier models do. If the copying in general has been accurate, all manuscripts will be fundamentally the same. But the differences between the manuscripts will be arranged such that later manuscripts will have more of the same changes than they have in common with earlier manuscripts.

Now, consider the possibility that two monks copy the same manuscript, and each introduces a different mistake. The two manuscripts are moved to separate monasteries, there to be copied—in isolation from each other—

over and over again by succeeding generations of monks. We now have two separate "lineages" of manuscripts. Within each lineage, all manuscripts have some unique peculiarities in common, and each succeeding manuscript has, in addition, more in common with its "descendants" than with the "ancestor." Both lineages converge at the ancestral manuscript, and the two lineages share all those features of the original that have not been modified by errors in copying over the ensuing centuries.

This analogy (though now rendered obsolete by scanners and copy machines) with biological evolution is entirely apt: manuscript historians predict that manuscripts copied in single, isolated monasteries are bound to share more errors in common than they will with other manuscripts copied elsewhere; thus, the history of manuscript transmission through the ages can be studied. Even when they don't know the exact chronological history of the copying process, manuscript historians can tell which of two manuscripts was copied first because of the distribution of errors in them.

The monk analogy applies in force to biological evolution: the very idea of evolution implies that each species will tend to have some features unique to itself, but each species must also share some similarities in structure or behavior with some other species. Furthermore, each group of similar species will share further features with other groups of species, and this common group must share features with still other groups. This pattern of sharing similarities with an ever widening array of biological forms must continue until all of life is linked up by sharing at least one similarity in common. And thus we arrive back at the first grand prediction of the very idea that life has evolved: the patterns of similarities in the organic world are arranged like a complex set of nested Chinese boxes.

We can go ourselves to nature and easily test this fundamental prediction. Does it work? Take any species—for example, the domestic dog *(Canis familiaris)*—and trace its relationships with other organisms. If this fundamental prediction of evolution is correct, there must be additional species that more or less resemble dogs—and of course there are: coyotes and various species of wolves. Somewhat more distant in terms of similarity, and hence relationship, we see that dogs are united with foxes and some extinct forms known only from fossils because all share some peculiar features of the middle ear. Members of this group (zoologists call them the Family Canidae) share other similarities (particularly of the ear region) with bears, raccoons, and weasels. In turn, all these creatures share carnassial teeth (in which the last upper premolar and first lower molar are bladelike and shear past each other like a pair of scissors) with cats, civets, and seals—the group zoologists call the Order Carnivora. Carnivores, it turns out, share three middle-ear bones, mammary glands, placental development, hair, and a host of other features with a number of other organisms, including humans. These organisms we call mammals. Mammals share with birds, lizards, snakes, and turtles an amniote egg, with its protective, enveloping tissues. Amniote animals share with frogs and salamanders the property of having four legs.

The kinship of dogdom widens as we see that some creatures, including all dogs, carnivores, mammals, and tetrapods, share backbones and other features with various sorts of animals we call fish. These animals, collectively, are the vertebrates. The circle widens to embrace progressively more groups: dogs, fungi, rose bushes, and amoebas have fundamentally similar (eukaryotic) cells. The eukaryotes are a massive, basic division of life, but they don't include bacteria and certain kinds of algae, for these are simpler yet, lacking the complex structures of the true eukaryotic cell. But bacteria and blue-green algae fall neatly into the fold when we look at the basic chemical constituents of all cells. RNA (ribonucleic acid), which copies the structure of the genetic material (DNA, or deoxyribonucleic acid) and sees to it that genetic information is translated into proteins, is found in all living things. And there, from a dog focal point, we have a quick rundown of the interrelatedness of all of life, as well as a capsule summary, at the same time, of the Linnaean hierarchy.

We started with dogs. We could have started with cats. The results would have quickly turned out the same (dogs and cats being so closely related). Had we started with ourselves, *Homo sapiens* (literally "wise mankind"), we would have found a nested grouping of ever widening similarities, starting with the great apes, then humans, apes, and Old World monkeys, and then that group (Anthropoidea) together with New World monkeys. Adding the lemurs and other prosimians, we would have found what zoologists call the Order Primates. From then on, the branch quickly melds with the dog line: primates, like carnivores, are mammals; mammals are amniotes; and so on.

Had we started with roses, the example would have taken longer to reach the same point of interrelatedness: roses share properties with various berries (Family Rosaceae), which share properties with other plants. All flowering plants are united by virtue of their shared mode of reproduction. All plants photosynthesize. Roses meet dogs only at the level of the Eukaryota. But the point is, they do meet. Thus this basic first prediction—the very notion that life has evolved—is confirmed: life, all life, as diverse as it is, is linked in a hierarchical arrangement of similarities. This must be so if evolution has occurred. This, indeed, is what we find.

More than 200 years of intense biological scrutiny leaves abundantly corroborated the fundamental idea that life has evolved. All organisms, including ones newly discovered on a daily basis, readily fall into this scheme. But it is even more important to see that the basic notion of evolution is inherently testable—hence inherently scientific. *Had we failed to find this nested pattern of similarities interlinking all forms of life, we would, as scientists, be forced by the rules of the game to reject the very notion of evolution.*

Ironically, creationists are no different from the rest of the citizenry of the United States in enjoying the myriad practical fruits of this first grand prediction of evolution: the application of predictions of similarity in essential fields of medicine and agriculture, not to mention the even more recent use of DNA testing in criminal law. For we routinely predict that, if we study a particular aspect of an organism—say, its digestive enzymes or the fine internal structure of its teeth—we will find exactly the same pattern of similarities already seen between this organism and others when biologists before us examined the hair, skulls, and fingernails. In other words, we predict that patterns of similarity of unexamined properties of organisms will agree with patterns already seen in more readily observable features. This must be so if evolution is a viable notion because life has had one single, coherent history.

This special notion of predictivity is vital to biomedical and agricultural research, which are the better-known areas of applied comparative biology. Not long ago, the news media carried a report entirely typical of the logic and structure of this kind of research. A doctor in Tennessee found that thiamine (one of the B vitamins) has a great positive effect in the treatment of lead poisoning. He performed his initial experiments on calves. Switching over to rats, he was disappointed to find the results weaker and less dramatic. Obviously, he told his interviewer, we should try thiamine on monkeys and apes suffering from lead poisoning. Why would he want to do that?

The good doctor was predicting that our own physiology (after all, it is treatment of lead poisoning in humans that motivated the research) would be more similar to the physiologies of monkeys and apes than to those of calves or rats. (Though because we share a more recent common ancestry with rats than with cows, I would predict that unfortunately his results with rats have greater implications for the treatment of lead poisoning in humans than do his results with calves.) Patterns of similarity seen in previous experience lead us to predict the existence of other, as yet unexamined, similarities. We expect the results of using thiamine as a treatment for lead poisoning in humans to be more similar to its effects on monkeys and apes than on either calves or rats, simply because we have known for centuries that we share more features with apes and monkeys than we share with any other sort of creature. It is this predictive feature of evolution, then, that underlies the entire rationale of biomedical experimentation on animals other than humans to assess the value and safety of various compounds to alleviate human ailments.

This simple prediction—that there is one grand pattern of similarity linking all life—doesn't *prove* evolution, but only because science proceeds by falsifying—*disproving*—statements we make about how the universe is structured and how it behaves. But we gain tremendous confidence in our statements if, after hundreds of years, everything we have devised to test an idea fails to falsify it. And so the failure of scientists to disprove evolution over the past 200 years of biological research means that the fundamental idea that life has evolved really is one of the few grand ideas of biology that has stood the test of time. This basic notion of evolution is thoroughly scientific in the strictest sense of the word, and as such is as highly corroborated and at least as powerful as the notion of gravity or the idea that the Earth is round, spins on its axis, and revolves around the sun. In the realm of science—and indeed in grander arenas of human knowledge and wisdom—evolution truly is triumphant.

# CHAPTER 3

# The Fossil Record
## Evidence of the Evolutionary History of Life

There is a second grand prediction that flows from the simple thesis that all life has descended from a single common ancestor—i.e., has evolved. Prediction 2: There should be a record of the evolutionary history of life preserved in the rock record, and that record should reveal a general sequence of progression from smaller, simpler forms of life up through the larger, more complex forms of life over long periods of time.

That there must be a history to life is obvious from the very idea of evolution: just as a family tree genealogy recreates the "begats" of remote ancestors—great grandparents, grandparents, and parents up to the current generation, all living in (overlapping) time periods from the past to the present—so too does the evolutionary notion of ancestry and descent among species imply the passage of time. Just how much time is necessary for evolution to have produced the full array of more than 10 million species, ranging from bacteria to tigers, is not intrinsically predictable (but it is, as we shall soon see, ascertainable through the techniques of geology and paleontology). But the very idea of evolution automatically invokes the passage of time every bit as much as does the notion of human family history.

Why would we predict there to be a general sequence from the simpler and usually smaller forms of life—single-celled organisms like bacteria—up to the multicellular creatures like us? Why not assume, instead, that everything arose virtually at once, or indeed progressed the other way down—from humans or elephants first, to bacteria last? Indeed, one of the stories of creation as recounted in Genesis calls for the creation of the Earth in just six days—and of all life in just two of those days—implying that the historical record of life should show everything appearing at exactly the same time, except we humans, who arrived a day later.

Despite recent claims of possible fossilized evidence of life brought down to the Earth in a Martian meteorite, we still have no definitive proof that life exists elsewhere in the universe. Until we have positive evidence to the contrary, in other words, we must assume (as Darwin did) that life arose here on Earth. Moreover, the proteins and the macromolecules that are the very foundation of living systems must have been formed from naturally occurring chemical precursors. Though experiments performed since the 1950s have been successful at synthesizing amino acids (the basic constituents of proteins) from such naturally occurring compounds as ammonia and methane, all the biochemical steps that led to the formation of the first organism—i.e., a system with the capacity to exist, buffered against a hostile environment, with the chemical wherewithal to reproduce (presumably with the chemical RNA first)—have yet to be deciphered. But clearly, if life arose naturally from prebiotic chemical constituents, the first living organisms must have been even simpler than the bacteria that abound around us and in a very real sense still run the world's ecosystems today. If life arose from nonlife, in other words, it would have to have been in a simple structural form. It could hardly have arisen, spontaneously, as an elephant.

Another way to look at this prediction is to think about the first grand prediction, which was discussed in the previous chapter. As we saw, the very concept of evolution yields the prediction that a grand pattern of similarity, through a complex set of nested groups arranged in hierarchical fashion, must in the end embrace absolutely every form of life known to exist—or to have existed in the past. And we saw that this prediction is abundantly verified by biological experience. The simplest forms of life on Earth right now are bacteria, which lack the truly complex cell structures found in other single-celled organisms, as well as in fungi, plants, and animals, but which nonetheless have biochemical pathways and, crucially, RNA and DNA every bit the same in basic structure as is found in the rest of the living world, including in our own bodies. In a very real sense, we can think of bacteria, the simplest known living things, as the least common denominator of all living systems, and as such in all probability remnants of the earliest forms of life ever to have graced our planet.

I now turn to a condensed review of the fossilized record of life—one that abundantly confirms our second prediction about the basic sequence of life on Earth, starting with primitive bacteria and bringing us up to the complex flowering plants and mammals (including humans) that were, as one would expect, much more recent arrivals on the planet. Much of creationist rhetoric is directed against the fossil record: creationists repeatedly claim that the fossil record of the history of life is *not* in accord with the general notion of evolution, and for this reason it is important to review the fundamentals of life's history as revealed not in the Bible, but in the rocks.

I have an additional motive in mind as I recount the story of life on Earth over the past 3.5 billion years. I have been a professional paleontologist for over three decades. Throughout my career I have been wresting from fossils clues about *how* life evolves—as we have seen, the second grand meaning of the phrase "theory of evolution," and the subject to which I turn in Chapter 4. I think of the fossil record—all 3.5 billion years of it, the last half billion chockablock full of wonderful complex animals and plants—as a series of "experiments" already performed by the evolutionary process. For I see hauntingly similar stories in the history of life—stories told by Cambrian trilobites, Permian corals, Jurassic dinosaurs, and even by our own ancestors in the rich fossil record of human evolution over the most recent 4 million years. The hauntingly similar histories of all these different kinds of creatures, spread out over a half billion years of geological time, form patterns in very much the same sort of way that a physicist's experiments in the cloud chamber of a linear accelerator reveal evanescent traces of the behavior of a subatomic particle—patterns that tell the physicist about the mass, charge, and other attributes of these otherwise unseeable bits of matter. My Devonian trilobites living 380 million years ago have left similar traces, as have all the other species that we are fortunate enough to have found in the fossil record of the evolution of life. And so I will—as we work our way up the history of life, with our eyes glued on our major prediction that life has had a rich history that confirms the general prediction that it has proceeded from simple to complex—also stop to point out clues that will help us when we turn in Chapter 4 to consider that second major aspect of the theory of evolution: our understanding of what factors cause life to evolve.

## In the Beginning . . .

We have a very good idea when the Earth was formed: about 4.65 billion years ago, most likely as a coalescence of gas, dust, and larger rocky particles. Three lines of evidence converge on this date: (1) the actual age of the oldest moon rocks collected, brought back to Earth, and analyzed; (2) the actual age of the oldest meteorites found (presumed to have formed in the asteroid belt at the same time the Earth was formed); and (3) the location at which the curve of dates of rocks on Earth projects back to "time zero." All these direct dates are ascertained from measurements of the ratio of remaining atoms (isotopes) that decay into daughter atoms at a statistically constant rate (see Chapter 5 for more on these techniques, and for a discussion of creationist attacks on atomic physics).

James Hutton (1726–1797), a Scottish physician and gentleman farmer who essentially founded the modern science of geology, predicted that we would never find the oldest rocks that had ever been on Earth. Hutton saw the evidence for a continual cycle of mountainous uplift, followed by erosion forming particles that would ultimately themselves be fused into rock, uplifted, and once again eroded. As Hutton (1795) put it, he saw "no vestige of a beginning, no prospect for an end." Hutton realized that there would be little chance for the Earth's oldest rocks to escape the forces that are continually tearing away and transforming rocks in our dynamic Earth. And he was right: the oldest rocks we have found to date—some 4.0-billion-year-old igneous rocks from Canada, fall a half billion years short of the actual age of the origin of the Earth.

Now, one of the most arresting facts I have ever learned is that life goes back as far in Earth history as we can possibly trace it. Fossils are found only in sedimentary rocks—i.e., rocks formed by the aggregation of particles of mud, silt, sand, or lime. Metamorphic and igneous rocks are simply formed in the wrong environments—especially under conditions of higher pressure and temperature—so whatever remains of organisms they may originally have contained is obliterated. But some of the oldest sedimentary rocks so far discovered—3.5-billion-year-old rocks in Australia—contain the fossilized remains of simple bacteria. And in even older rocks (3.8 billion years old), geochemists have found forms of carbon atoms (isotopes) considered by biochemists as a fingerprint of the presence of life. In other words, in the very oldest rocks that stand a chance of showing signs of life, we find those signs—those vestiges—of life. Life is intrinsic to the Earth!

One of creationism's fondest lines of attack on evolution is the claim that science has not yet fully solved the riddle of how life originated. But, as Charles Darwin himself proclaimed in his *On the Origin of Species*, the question of the origin of life is separate from the issue of what happened to life after it did arise. The earliest fossils we have are of very simple, mostly rod-shaped bacteria—far advanced over the simplest of molecules capable of self-replication that must have constituted the earliest forms of life.

But consider what a bacterium is compared to all other forms of life. Bacteria are prokaryotes: they lack the nuclei typical of the cells of all other organisms—from single-celled creatures (the protoctists) such as amoebas, up through the multicellular plants, fungi, and animals. The nuclei of the cells of these more complex organisms house most of their DNA—the genetic material that serves the dual functions of running the basic machinery essential for staying alive and replicating itself in the reproductive process—allowing, in other words, the equally essential process of reproduction of the organism to occur. DNA is the template from which RNA copies genetic information, in turn taking it to the ribosomes, where amino acids are assembled according to that information to form proteins. It is the proteins that run the cell, catalyzing chemical reactions and forming the building blocks of the cell itself.

These more complex organisms (i.e., the single-celled protoctists, plants, fungi, and animals) almost always have not one, but two, copies of each gene organized along pairs of chromosomes in the nucleus of each cell. Bacteria have a simpler setup: since there is no nucleus, the single-stranded genetic material is simply disseminated in strands in the body of the organism. Thus, complex as they are, bacteria are fundamentally far simpler than all other organisms that have ever existed on Earth.[1]

Thus, evolution predicts that the simplest kinds of organisms we know should be the oldest ones we find in the fossil record. And that is exactly what we do find.

In addition, at the very earliest stages of life's existence on Earth, we meet up with a theme—a recurrent pattern—that is utterly typical of life's entire 3.8-billion-year history, a theme that will figure prominently in my account of the evolutionary process in the following chapter: almost without fail, whether we are considering ancient bacteria 3.8 billion years old or the evolution of the human lineage in the last 4 million years, we find that significant events in life's history are correlated with significant events in the physical history of the Earth's atmosphere, oceans, and lands. For example, there is mounting geochemical evidence that the Earth was bombarded by many extraterrestrial objects—apparently comets—at the same time that the moon was pockmarked by its own bombardment 3.8 billion years ago. Some geologists have speculated that this tremendous comet bombardment supplied most of the water we now have on the Earth's surface—water so essential to life, trapped here by our gravitational field, but lost on the atmosphere-lacking moon, which lacks sufficient mass to keep gases and water molecules around it. Others think that primitive forms of life itself[2] might have been brought to Earth during this bombardment. Comets, after all, are "dirty snowballs," and the organic chemical compounds that are the basic constituents of life are common in outer space.

Still others think that whatever life might have existed on Earth *before* this monumental cometary bombardment 3.8 billion years ago may well have become extinct—meaning that life may well have developed not once, but twice, on Earth. Clearly we need to know more about the details of early Earth history to resolve these questions. But the very fact that these issues exist shows two very important things about science:

First, the great amount we have learned about both the very earliest stages of the Earth and life on Earth allows us to pose the questions in the first place. When I was in graduate school in the 1960s, pre-Cambrian paleontology was in its infancy. When Darwin was writing in the mid-nineteenth century, paleontologists knew absolutely nothing about the history of life before the appearance of the earliest forms of animal life (which we now understand happened about 535 million years ago). Now we see an increasingly rich early fossil record—one restricted to bacteria for at least its first 1 *billion* years—but rich enough for us to get some definite ideas of what the earliest phases of life's evolution were like.

Second, we see the creative side of the scientific mind at play. New questions arise from the analysis of data, and of repeated patterns in the history of life. The reason why anyone would speculate that the massive cometary bombardment 3.8 billion years ago caused an extinction of the Earth's earliest life forms is simply that there is now consensus in the scientific world that much later—65 million years ago—a smaller-scale cometary bombardment killed off the last of the dinosaurs, as well as many other marine and terrestrial species. We now understand that extinction caused by physical disaster is a fact of the history of life; if it could happen 65 million years ago, why not 3.8 billion years ago?

Thus science is a form of human knowledge, and like every other branch of human knowledge, it grows. We have definitive rules—canons of judgment—by which we decide which ideas are probably valid and which are not. We need good hard evidence, detectable by our senses. Obviously, not every idea of those I have mentioned about what happened to life 3.8 billion years ago on Earth can possibly be true at one and the same time. The evidence to date is tantalizing but scant, so we leave the ideas out on the table, certain that one day we'll garner even more evidence—geological, molecular, fossil—to focus the picture even more clearly, and to narrow the range of possible answers. We may even come up with still newer ideas. As we shall see in later chapters, creationists like to call the changeability of scientific conclusions the Achilles' heel of science, but the growth of knowledge—scientific or otherwise—absolutely depends on keeping a collective mind open to all (rational!) possibilities, weeding through them as the evidence mounts, before a consensus on what really happened is reached. And then it is on to the next exciting problem!

Bacteria still rule the Earth. We tend to overlook that fact, since for all intents and purposes bacteria are invisible. We know them mainly because they cause infections. But they do so many other things as well. One small example: were it not for bacteria living in the hindguts of termites, most cellulose—especially in arid regions of the Tropics—simply would not be broken down. Without decay, the world would soon be literally choked with dead wood. Some bacteria (the so-called blue-green algae) photosynthesize, among other things adding to the world's supply of oxygen. Others are vital for the cycling of nitrogen, carbon, and sulfur through the world's ecosystems.

We also overlook bacteria because of the old-fashioned tendency to call the earliest stages of life the Age of Bacteria, followed by the Age of Invertebrates, the Age of Fishes, the Age of Reptiles, the Age of Dinosaurs, the Age of Mammals, and then finally the Age of Mankind. It has been traditional in evolutionary circles to stress the progressive aspects of life's history, so when something new comes along (e.g., complex animal life), there has been a tendency to focus on the new, forgetting about the earlier forms of life that are still very much present. It is abundantly evident, though, that the evolution of new, including more complex, forms of life by no means implies that the earlier, simpler, more primitive forms are thereby somehow rendered obsolete. They do not, as a rule, become extinct, and their presence is as important in the world's ecosystems today as it ever was.

Evolution, as my colleague Stephen Jay Gould likes to say, is much more like a richly branching bush than like a simple ladder of progressive change. When new, often more complex life forms evolve—finding a role, and thereby surviving, but by no means supplanting the forms of life that preceded them—they enrich life's diversity by being added to the mix of what was already there. Thus the second grand prediction of life—that more complex forms of life necessarily evolve from, and thus come later than, more primitive forms—does *not* include (as some creationists have claimed) the idea that later, more "advanced" forms are necessarily superior to, or drive to extinction, earlier, more primitive forms of life.

After bacteria, the next major step in life's history was the evolution of the eukaryotic cell—the complex type of cell with its own discrete nucleus housing pairs of gene-bearing chromosomes that is shared by many singled-celled organisms (such as amoebas), as well as all fungi, plants, and animals. The oldest known fossils of eukaryotes are single cells recovered from sediments approximately 2.2 billion years old. In the summer of 1999, scientists reported finding chemical traces typical of eukaryotic organisms that are even older: some 2.7 billion years old, though to date, no actual fossils of eukaryotic organisms that old have been found. Note that, just as we would predict, these earliest eukaryotic organisms are the remains of simple, single-celled protoctists—i.e., not more complex, multicellular plants, fungi, or animals. Thus, just as we would expect, the next major step in the evolutionary history of life was to go from the relatively simple structure of a bacterial cell to the more complex structures of the eukaryotic cell—but still at the level of single-celled organisms, and *not* at the even more complex level of organization of multicellular organisms (fungi, plants, and animals).

Biologist Lynn Margulis has proposed an explanation for this great evolutionary step—one that, like the notion of a round Earth, was greeted with disbelief and even derision as a crackpot idea when she first proposed it in 1967. But her idea is so powerful, with so much corroborating evidence supporting it, that by the 1980s and 1990s, it had become common knowledge—a powerful, abundantly confirmed hypothesis that itself has virtually turned into a fact. Margulis proposed that the eukaryotic cell arose as a symbiotic fusion of two (or more) different kinds of bacterial cells. "Symbiosis" literally means two forms of life living together in close association for each other's mutual benefit. Generally, the association is so close and intense that the forms simply must live together, and sometimes it is so close that the two different life forms seem to be one single organism. Lichens, for example, are symbiotic associations of algae and fungi.

The cells of eukaryotes contain organelles—miniature systems outside the nucleus, embedded in the cytoplasm and performing various metabolic roles. The ribosomes, for example, are the sites for assembly of proteins, controlled by transfer RNA that has copied the instructional sequences from the DNA of the cell's nucleus. Mitochondria, another kind of organelle, are the sites where respiration (chemical extraction of energy from food) occurs. In plants, cells are additionally equipped with chloroplasts, organelles in which the chemical reactions of photosynthesis occur. One of the more surprising results of molecular biology was that *both mitochondria and chloroplasts have their own DNA*—single-stranded, very much the way DNA occurs in bacteria. Margulis proposed that mitochondria and chloroplasts were originally free-living bacteria that invaded other bacterial cells and, in effect, specialized in energy production while other parts of the newly evolved eukaryotic cell specialized in other tasks. The other parts of the eukaryotic cell may also have arisen as separate bacterial invasions, though the evidence is by far the most persuasive for the chloroplasts and mitochondria because they still have their own separate complements of DNA and RNA. And, sure enough, mitochondrial DNA replicates as the rest of the cell divides.

Biologists quickly came to accept Margulis's thesis on the origin of chloroplasts and mitochondria once it became clear that the DNA of these organelles is always present and bears no resemblance or connections with the DNA of the nuclei of eukaryotic cells. There is simply no other rational explanation for the arrangement of DNA within all eukaryotic cells. Cellular fusion—evolutionary symbiosis taken to its final conclusion—accounts for the basic division of labor of the organelles of the eukaryotic cell.

What prompted this major evolutionary advance: the origin of the eukaryotes? Though here again we cannot be sure, recent work by Caltech geologist Joseph Kirschvink and his colleagues strongly suggests that the world underwent a truly bizarre and extreme climatic event around the time eukaryotic life evolved. Kirschvink calls it "snowball Earth": The geological evidence is persuasive that, around 2.2 billion years ago, continental glaciers started growing as extensions of the polar ice caps. But, unlike more recent glacial events, Kirschvink has found that these massive sheets of ice reached all the way into the Tropics—and probably right down to the equator 2.2 billion years ago.

Did snowball Earth somehow *cause* the evolution of the eukaryotes sometime over 2 billion years ago? We simply do not know enough about either the physical or the biological events back then to be sure. But it is a tantalizing possibility: the approximate correlation in timing of a major step in life's evolutionary history with one of the most severe climatic episodes ever to occur on Earth strongly hints of a connection between these events.

Once the single-celled eukaryotes had evolved, there were no major advances in evolutionary history for some 1.5 billion years. Though undoubtedly the single-celled protoctists diversified, their fossils are difficult to find, and we have relatively few samples over this 1.5-billion-year interval to give us insights into the nature of eukaryotic single-celled life during this period. Life remained small and relatively simple; the larger and more complex animals, plants, and fungi had yet to appear.

## The Cambrian Explosion

Creationists are very fond of the Cambrian explosion, the relatively abrupt appearance of complex animal life that marks the beginnings of a rich and dense fossil record. Creationists say that the second grand prediction of evolution—that life has evolved in an orderly fashion, with simpler forms preceding more complex—is violated, even downright falsified, by the early Cambrian fossil record. Paleontologists, on the other hand, see this explosion as a fascinating example of the phenomenal speed at which evolution can work. What is this Cambrian explosion exactly, and how well does it conform to our prediction that simpler forms of life precede the more complex?

As geologists in the late eighteenth and early nineteenth centuries began the job of mapping the sedimentary rocks of Europe and North America in earnest, they saw that all the rocks lying below what they called the Cambrian System ("Cambria" was the Roman name for Wales, where Reverend Adam Sedgwick [1785–1873] first studied and named rocks of this age) seemed devoid of fossils. In contrast, from the Cambrian on up, fossils of complex life—first marine invertebrate animals like corals, brachiopods, mollusks, and trilobites; and later vertebrate life (fishes); and even later, invertebrates (insects), vertebrates (amphibians; later, reptiles; still later, birds and mammals), and plants on land—became consistently abundant.

Some of the oldest of these fossils in the Cambrian are of trilobites, the most primitive group of the Phylum Arthropoda (literally "jointed-legged ones"), which also includes crustaceans (crabs, barnacles, shrimp), millipedes, horseshoe crabs, spiders, and insects. That arthropods, as all biologists agree, are much more complicated animals than, say, sponges or corals presented a puzzle: Why are some of the oldest forms of animal life that are found in the fossil record also some of the most complex?

Thus was born the riddle of the Cambrian explosion—made even more puzzling after Darwin convinced the scientific world that life had indeed evolved in an orderly fashion. Paleontologists speculated that there must be an interval of time missing, or that all the traces of even more ancient life—traces that would have recorded the transition from single-celled eukaryotes to simple animals such as sponges or corals—had been obliterated by erosion and metamorphism.

Though it is true that the vast majority of so-called Precambrian rocks are indeed metamorphic or igneous, nonetheless there are sequences of sedimentary rock that lie below fossiliferous Cambrian rocks. These really *ought* to produce forerunners and precursors to the abundantly preserved hard-shelled invertebrates that show up in such profusion in Lower Cambrian rocks. And sure enough, diligent searching and collecting, primarily in the latter half of the twentieth century, has—as we would predict from the simple idea of evolution—shown that there was complex life before the Cambrian trilobites, brachiopods, mollusks, and sponges burst upon the scene.

Geologists have recently discovered that perhaps as many as three or four more prodigious "snowball Earth" glaciations occurred over a 200-million-year interval beginning about 800 million years ago. We find, for the very first time, the fossilized remains of large organisms that lived right after the last of these stupendous glacial events had subsided some 600 million years ago. Best known from the Ediacara Hills of the Flinders Range of Australia, elements of this so-called Ediacaran fauna of the Vendian Period (600–540 million years ago) have been found in such far-flung places as Newfoundland, the Charnwood Forest of England, and Namibia in southern Africa. The fossils come in a variety of shapes and sizes,[3] and frankly, the paleontological world has yet to reach complete consensus on what these Ediacaran fossils really are: Are they true animals, primitive versions of later forms of life? Are they instead lichens, as has been suggested? Or are they a form of life that flourished and then became utterly extinct—leaving it very difficult for us to figure out exactly what manner of beast they were?

On the other hand, many of the Ediacaran fossils are very similar to well-known forms of invertebrate life. *Spriggina*, for example, looks to most of us paleontologists very much like a segmented worm and, with its rather well developed head region, perhaps a forerunner to Cambrian trilobites. Likewise, I see no problem in relating others—indeed the vast majority—of the Ediacaran fauna to various groups of cnidarians, also known as coelenterates (especially sea pens—soft-bodied relatives of corals), and it is well worth noting that cnidarians (which include corals, anemones, and jellyfish, which lack true organ systems) are among the simplest, most primitive forms of animal life on the planet. Still others (e.g., *Tribrachidium*) remind me forcibly of echinoderms (which include starfish and sea urchins, especially the later-occurring (in the Cambrian Period) edrioasteroids, a now long-extinct group of echinoderms. Though some paleontologists still disagree, for my money I think we will eventually conclude that many, if not all, of these fossils Ediacaran are closely related to the corals, echinoderms, worms, and arthropods that they seem to resemble.

Thus, predictably, we have begun to fill in the gaps; we now know that the advent of complex multicellular animal life did not occur overnight (nor in a single biblical day), but rather took place in a succession of events spanning 160 million years. But what of that relatively sudden, abrupt appearance of trilobites and other complex forms of animal life at the base of the Cambrian?

Recent research, initially mostly on fossil faunas in Siberia, has led to the discovery of a variety of generally small shell-like forms that, like their Ediacaran predecessors, are difficult to assign to well-known groups with utter certainty. They look, for the most part, like mollusks—snails, clams, and cephalopods being later, more familiar molluscan examples. But then, sometime around 540 million years ago, we begin to find an array of fossil remains of what were quickly to become some of the dominant elements of marine life for hundreds of millions of years. Trilobites, for example, make their appearance, as do brachiopods (bivalved creatures unrelated to the bivalved clams; brachiopods still exist in modern seas), as well as early mollusks and calcareous spongelike creatures known as archaeocyathids.

First impressions—including those of the early geologists—see this as an instantaneous explosion of a vast assortment of invertebrate life forms. We now know that trilobites and other sorts of Cambrian life did not show up absolutely all at the same time all over the world; rather, it apparently took a good 10 million years for the familiar faunas of the lowermost Cambrian rocks to become fully established.

I have walked up a dry creek bed in the White-Inyo Range on the California-Nevada border, carefully observing and collecting while climbing a rock sequence near the very base of the Cambrian. I'll never forget finding exactly what my geologist guide had told me I would: that, in the lowest (therefore oldest) sediments, I would see only trace fossils—meaning the sedimentary structures formed as animals moved through, or walked on top of, the bottom muds of that ancient sea. Some of these traces were clearly made by animals as complex as trilobites. But where were their skeletal remains, which as we walked up the hill, all of a sudden became common and easy to find and collect? Either there was something peculiar about the chemical environment when the lower sediments were accumulating—something that prevented skeletal material from surviving long enough to become buried and eventually preserved as a fossil—or the skeletons were too thin and delicate to make it all the way to the burial-and-preservation stage.

I simply don't know, in this particular case. But it is clear that the ability to grow to large sizes and to secrete hard material like calcium carbonate to build a tough external skeleton—the sort of skeleton that readily becomes fossilized—was severely hampered by the chemistry of the ancient seas. One particularly attractive notion (the Berkner-Marshall hypothesis), well supported by geochemical measurements, is that there simply was not enough oxygen dissolved in seawater until about 540 million years ago—enough, that is, to support the metabolic requirements of large-bodied, multicellular organisms, including their ability to secrete hard external skeletons.

So it seems likely that the ancestors of the major groups of invertebrates had already diversified to some extent *before* they had grown very large and had developed external exoskeletons. Indeed, that is what I think the Ediacaran organisms are telling us (though some of these did reach respectable sizes, albeit apparently without forming truly well mineralized exoskeletons). Recall, too, that it was the coral-like forms that dominated life in the Ediacaran seas.

The invertebrate phyla alive today—and alive since Cambrian times—show a nice evolutionary gradation from simple to complex. Simplest are the sponges, which lack true tissues, having only a few different types of cells. The cnidarians (corals, anemones, various sorts of jellyfish and polyps) come next in the spectrum of complexity: these animals have two distinct layers of tissues in their bodies, though they lack true organs. More complex are such phyla as the flatworms, which have eyes and excretory organs, for example, but lack segmentation or a true body cavity. Still more complex are the brachiopods and bryozoans; the segmented worms, mollusks, and arthropods; and the echinoderms and chordates (the latter is our own group, but some chordates are still considered invertebrates[4]).

Once inside the Cambrian, we find no direct succession through time of the simplest invertebrates up through the most complex. There is no sequence by which sponges enter first, followed by corals, then flatworms (never, in any case, found as fossils), then brachiopods, worms, trilobites (arthropods), and so on. There are two reasons for this lack of succession, implicit in what we have seen so far. First, it is pretty clear that these groups had already diversified—evolved—from each other sometime during the Vendian— prior, that is, to the time when many of these groups, roughly at the same time, presumably for geochemical reasons, attained the ability to secrete salts into their outer coverings, giving them hardened skeletons for the first time. That, I think, is one of the main lessons of the Ediacaran fossils.

Second, no biologist would ever propose that sponges gave rise to corals, which gave rise to flatworms, which gave rise to mollusks, which gave rise to . . . , and so on ad infinitum—like the long list of "begats" in Genesis. For if you look at the small larvae of all these groups, similarities quickly become evident that have suggested to biologists over the years how the actual ancestors in this grand evolutionary radiation really looked. Larvae are small, and they are soft-bodied—just as are some of the fossils we are only now beginning to recover from Vendian sediments. The hard-bodied invertebrates that began showing up in profusion some 540 million years ago were already well differentiated and diversified, and they were descended from much smaller, soft-bodied forms that had evolved sometime in the 100 million or so years after the last-documented snowball Earth—well before, that is, the base of the Cambrian.

In the old days my predecessors predicted the existence of what they called the Lipalian interval—an interval of time just before the Cambrian during which the great radiation of invertebrate animal phyla took place. The problem was, they couldn't find it. Now we've found it, and we have dubbed it instead the Vendian. As paleontological techniques become ever sharper, we almost certainly will find more direct evidence of the pattern of this great invertebrate diversification—one that, if it ever becomes good enough, will show that the less complex groups (such as the ancestral sponges and cnidarians) actually did precede, even if not by much, the more complex groups like mollusks and arthropods.

### After the Radiation: The Pulse and Pace of Evolutionary History for the Last 540 Million Years

Once complex marine animal life finally became established some 540 million years ago, and all the major groups except the vertebrates (in the form of primitive fish[5]) had appeared, patterns of evolution become far easier to trace in the rocks. Once again we find a simple agreement between the fossil record and evolution's second grand prediction—that simpler, more primitive forms precede their evolutionary, more complex, descendants. We also see the beginnings of a repeated set of patterns to life's evolution that are still going on around us today—patterns that have much to tell us about *how* life evolves.

Let us consider the pulse and pace of evolution—the typical evolutionary patterns—that we begin picking up in the Cambrian Period, at the beginning of the Paleozoic Era. Five hundred million years ago, life was still restricted to the seas (at least as far as we know; bacteria were undoubtedly also living in fresh waters and terrestrial environments, but the record of these ancient environments is poor to nonexistent). Marine life was both plentiful and richly diverse—with trilobites, brachiopods, the spongelike archaeocyathids, primitive echinoderms (starfish relatives), and primitive mollusks especially common. In one justly famous quarry in British Columbia, an astonishing array of soft-bodied animals discovered in the Burgess Shale[6] has shown us how incredibly diverse early life in Cambrian seas was, also showing us that most marine deposits, no matter how richly fossiliferous, contain the remains of only easily preserved organisms—i.e., those with hardened skeletal parts.

Paleontologist Allison R. ("Pete") Palmer, formerly of the State University of New York at Stony Brook, and since retired from the U.S. Geological Survey, has documented a succession of faunas in Cambrian rocks of the western United States. Palmer was impressed by the tendency of entire groups of trilobites to flourish, and in particular to evolve many different species relatively quickly—species that would persist for millions of years before sudden environmental change (Palmer thought it was the rising of sea level) would disrupt these nearshore shallow-water ecosystems, abruptly driving many species extinct. Palmer called each of these successive evolutionary and ecological episodes a biomere.[7]

Palmer's work (primarily in the 1960s and 1970s) was the first of its kind in the modern era to expose the fundamental nature of the organization of the fossil record for the last half billion years of life on the planet. The pattern of rapid evolutionary diversification, followed by long intervals of great ecosystem stability with very little evolutionary change detectable within species, followed (after millions of years) by physical disruption of ecosystems and loss of many species to extinction, and then followed by another burst of rapid evolutionary change, is absolutely typical of evolutionary patterns on land and sea for as long as complex animal (and, later, plant) life has been on Earth.

Indeed, this pattern can be found virtually everywhere in the fossil record of the history of life. For example, paleontologists Carlton Brett and Gordon Baird[8] have documented eight successive intervals very reminiscent of Palmer's biomeres—but this time from the Middle Paleozoic of the Appalachian region of eastern North America. Each of their recognized intervals lasts 5–7 million years—an aggregate total time of some 45–55 million years. Brett and colleagues estimate that between 70 and 85 percent of the species are present throughout their respective intervals, and only some 20 percent, again on average, make it through to the next succeeding interval. Their name for this type of repeated pattern in the history of life is coordinated stasis.

There are many other examples: dinosaur faunas of the Mesozoic, mammalian faunas of the Cenozoic, marine Cenozoic faunas—even the great Rift Valley sequences of eastern and southern Africa, where our own evolutionary drama was played out—all show exactly the same pattern: (1) great ecological stability, with great evolutionary stability of animal and plant species, followed by (2) physical disruption of ecosystems, with high rates of species extinctions, followed by (3) rapid evolutionary proliferation of descendant species. But we should not lose sight of the fact that the new species replacing the old ones are always different, equipped with newly evolved adaptations. Nowhere is this characteristic sequence of events more obvious than in the details of human evolution in Africa, which fit this pattern exactly, as we shall shortly see. In the African case, the same pattern goes by the name "turnover pulse," coined by paleontologist Elisabeth Vrba, who was the first to pinpoint the importance of this pattern in the evolution of African mammals, including our own hominid ancestors.

Here is another pattern that has great potential significance for understanding *how* the evolutionary process works—the subject of the next chapter: during the long intervals of time between environmental disruption, extinction, and the rapid subsequent development of new species, ecosystems and species themselves are remarkably stable. Little or no evolutionary change accumulates in most species during these periods of quiescence—a phenomenon not greatly remarked on by biologists until my colleague Stephen Jay Gould and I discussed it in the 1970s, calling it stasis.

All the great biologists and geologists prior to Darwin were, in some sense at least, creationists. Evolution had yet to gain scientific credibility up through the first half of the nineteenth century, and most educated Europeans were religious as a matter of course. In the early 1800s, the great French biologist-geologist-paleontologist Baron Georges Cuvier (1769–1832) recognized no fewer than 32 separate divisions of the fossil record of life—in a crude, beginning way, seeing the history of life very much as we see it now—as a succession of extinctions followed by proliferation of new ecosystems with new species to fill the roles vacated by their extinct predecessors. Cuvier thought that the proliferations were independent creations by God (unlike modern-day creationists, who follow Genesis literally and see only a single creative episode for all of life); it is the triumph of the Darwinian-inspired evolutionary perspective that allows us to see them for what they really are: evolutionary responses to ecological change. And it was Cuvier who, despite creationist interpretation, was among the first to draw attention to these all-important patterns of stability and what he called revolutions.

Global mass extinctions leave far more devastation in their wake than the regional pattern variously dubbed biomeres, coordinated stasis, and turnover pulse—and produce far greater evolutionary results. The greatest extinction (so far!) to have hit Earth's biota took out at least 70 percent— and possibly as much as 96 percent—of all species living on Earth, according to paleontologist David M. Raup, formerly of the University of Chicago. That was some 245 million years ago, at the end of the Permian Period— an event that changed the complexion of life so deeply that early paleontologists (long before Darwin) used it to divide the Paleozoic Era (Ancient Life) from the Mesozoic Era (Middle Life).

But it is not just individual species that succumb to extinction in these massive, global extinction crises. When global extinction events strike, they tend to take out so many species that entire families, orders, and even classes— the larger units of the Linnaean hierarchy—disappear. And on the other side of the line, when whatever caused the extinction event itself passes, life rebounds, and entire new families, orders, and even classes of animal and plant life become established. Let us look at some examples of these larger-scale evolutionary changes.

Ammonoids are coiled (almost always, though there are exceptions), externally shelled cephalopod mollusks—close relatives to the living pearly nautilus of the South Pacific. (Other living cephalopods include squids and octopi.) Ammonoids evolved in the Devonian Period from nautiloid ancestors; they became extinct at the end of the Mesozoic Era, in the fifth great mass extinction, which occurred some 65 million years ago.

The ammonoid shell, like that of the pearly nautilus, is divided by a series of partitions. These partitions (called septa) intersect the outside shell wall and leave a characteristic pattern of hills and valleys that is especially visible on fossils in which the outer shell has been worn away. It is these suture patterns that show so much evolutionary change during the course of ammonoid history. And though there are some collateral groups of ammonoids, by far the dominant groups were the Paleozoic goniatites, the

Triassic ceratites, and the Jurassic and Cretaceous ammonites. The earliest ammonoids, the goniatites, had the simplest sutures, expressed as smooth lines arrayed in peaks and valleys (or "saddles" and "lobes," to use the terms of ammonoid paleontology). The later ceratites had secondary crinkles on the lobes but not on the saddles. The last great group—the true ammonites—had crinkles on both the lobes and the saddles. [see Figure 1]

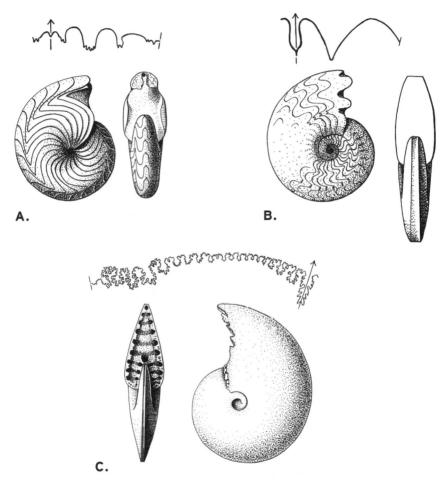

A.

B.

C.

**FIGURE 1** The three basic kinds of ammonoid suture patterns, showing suture pattern on the side of the shell, a view of a septum, and the outline of the suture pattern. **A.** The Mississippian goniatite *Imitoceras.* **B.** The Triassic ceratite *Meekoceras.* **C.** The ammonite *Placenticeras.* From Easton, 1960, pp. 441, 448, and 461.

Now here is progressive evolution in the old, classic sense, for anatomical complexity increases through time in ammonoid history. But consider the timing and the circumstances behind these events: all but one genus (small group of species) of goniatites became extinct at the Permo-Triassic extinction event at the end of the Paleozoic. But those few related goniatite species gave rise to the first of the ceratites, which rapidly radiated in an explosive evolutionary proliferation of species. Soon, as Triassic ecosystems began to regain the normal look typical of marine ecosystems throughout the ages, the ceratites were well established as the dominant group of ammonoid cephalopods.

The very same thing happened after the next major mass extinction event, at the close of the Triassic Period. The ceratites became extinct, but from one surviving ammonoid lineage the entire vast array of the still more complexly sutured ammonites evolved. And they were to proliferate and thrive until the next extinction—the one at the end of the Cretaceous that ushered in the Cenozoic Era—finally drove the last of the ammonoids to extinction. The evolution of entire large-scale groups is often controlled by larger-scale versions of the pattern of extinction followed by evolution that we have already seen is typical of smaller-scale evolutionary events throughout the history of life.

Another example that illustrates such large-scale evolutionary patterns involves corals. Corals are related to sea anemones, jellyfish, and other soft-bodied cnidarians (Phylum Cnidaria). Corals, though, have a hard external skeleton made of calcium carbonate, a feature that makes them commonly preserved elements of the marine fossil record from the Paleozoic on up to modern times. Corals became common in Paleozoic seas in the Lower Middle Ordovician, as limy seafloor environments became widespread over the interiors of the great continental landmasses, especially Asia, Europe, and North America. Some earlier corals, however, are known, and some in the Cambrian of Australia look very like the modern corals we have today—the Scleractinia, also known as hexacorals for their sixfold symmetry of septa supporting the soft tissue of the coral polyp itself.

But the dominant corals of the Paleozoic Era—corals that often formed massive reefs—had a fourfold symmetry to their internal septal arrangements.

Moreover, their skeletons were formed from the mineral calcite, the more stable atomic arrangement of calcium carbonate than the delicate aragonite form of calcium carbonate. The big problem in coral paleontology always was, How did the modern scleractinian corals—found first in Lower Triassic rocks—evolve from the Paleozoic corals, which were last seen in the Upper Permian? How did the mineralogical change (calcite to aragonite) occur? And how, after hundreds of millions of years of fourfold symmetry, did six-sided symmetry evolve? [see Figure 2]

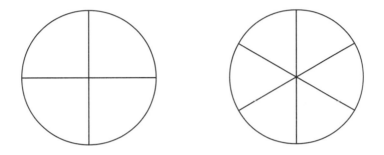

**FIGURE 2** Coral: Four-fold symmetry vs. Six-fold symmetry.

The answers to these puzzles are a bit different from what we just saw in ammonoid evolution; in that example, evolutionary advances in complexity occurred in small groups of species that somehow managed to survive when most of the older ammonoids became extinct. The corals present a somewhat different evolutionary story: We now understand that scleractinian corals did not evolve directly from the Paleozoic rugosans or tabulates. We now understand, instead, that the nearest relatives of scleractinian corals are actually the naked sea anemones (which show the same pattern of six-sided internal symmetry). It now seems that the early Cambrian corals were part of a sixfold-symmetry lineage that persisted—largely as naked anemones that have left only a few fossil traces in the Paleozoic, but a lineage that was nonetheless always there, throughout the Paleozoic. When the older forms of corals died out in the great Permian mass extinction event, one branch of the sixfold-symmetry (anemone) lineage independently acquired the ability to secrete calcium carbonate—albeit in the form of aragonite, not calcite.

Thus evolution seems very much to involve the reinvention of the ecological wheel: when the Paleozoic corals became extinct, a related branch of cnidarians independently evolved the ability to secrete calcium carbonate, thereby evolving into what can only be described as the general coral niche, and producing so many species that zoologists classify them as the Order Scleractinia. In that sense, the Orders Rugosa and Tabulata of the Paleozoic were replaced by the Order Scleractinia. But note, too, that the Scleractinia almost certainly would not have evolved had not the Rugosa and Tabulata been driven to extinction. This is the evolutionary phenomenon that paleo-biologist Stephen Jay Gould has called contingency: had those Paleozoic corals not succumbed to extinction, the corals that make up today's Great Barrier Reef in Australia would never have evolved.

Here's an even more graphic example: In the Upper Triassic, a little over 200 million years ago, the earliest mammals and the earliest dinosaurs appeared for the first time. The dinosaurs clearly had evolved from the very similar, but more primitive, thecodont reptiles living in the Middle Triassic. The mammals had evolved from a different branch of the reptiles: the synapsids, reptiles with but a single hole in the side of the head (as opposed to the two holes characteristic of snakes and lizards, crocodilians, birds, dinosaurs, and still other diapsid reptile groups). Indeed, the evolution of mammals from the "mammal-like" reptiles (therapsids) is one of the best-documented examples of macroevolution ever encountered in the fossil record. Especially known from South Africa's great Karroo System of sediments spanning the Permian and Triassic Periods, the sequence of fossils progresses upward from creatures with undifferentiated tooth rows and a single middle ear bone right on up through very mammalian-looking animals with teeth differentiated into incisors, canines, and molars, and with three middle ear bones.[9] The Karroo sequence—like nearly all other segments of the fossil record of life's history, is separated into distinct biomere-like zones, and the changes in early mammalian evolution take place in the same sort of pattern of stability followed by extinction and then evolutionary advance that is seen in all other parts of life's evolutionary story.

But it was the dinosaurs, not the mammals, that evolved into a rich array of small, medium, and large species—some herbivores, some scavengers, and some, of course, carnivores. In the great extinction event that closed the Triassic Period, the dinosaurs were cut back severely, but they did not succumb completely to extinction. They bounced right back—in an evolutionary sense—and once again held, as it were, most of the important jobs in the world's terrestrial ecosystems. Though they suffered vicissitudes throughout their 150-million-year reign, nonetheless it was always the dinosaurs, not the mammals, that were able to evolve more species and to diversify, playing a wide variety of roles in Earth's ecosystems.

No one knows why it was always the dinosaurs, and never the mammals, that were able to evolve such a wide assortment of body sizes and ecological types. But finally, at the end of the Cretaceous Period 65 million years ago, the last of the dinosaurs finally did succumb to the environmental horrors thrown up by the collision between Earth and one or more comets—and, the thinking now goes, by the correlated outpouring of vast columns of lava on the other side of Earth, in peninsular India.

So, after a long reign in which evolution repeatedly refreshed their diversity, suddenly the dinosaurs were gone. The mammals, which had diversified into several distinct lineages, but which had not developed anything like the diverse array of ecological types that the dinosaurs had, managed to hang on through the mass extinction event at the Cretaceous-Tertiary boundary. In the words of famed Harvard paleontologist Alfred Sherwood Romer (1894–1973), mammals had remained "the rats of the Mesozoic" (personal communication), definitely playing second fiddle to the dinosaurs and the other ruling reptiles.

It took only a few million years—the characteristic lag seen after all major mass extinction events—before the mammals suddenly burst forth in a blaze of evolutionary activity. All of a sudden, huge lumbering herbivorous mammals appeared on the Paleocene landscape; by the Eocene, recognizable ancestors of all the modern groups of mammals had evolved. The death of the dinosaurs did not cause mammals to evolve; they had already evolved long since. But it did bring about the evolution of mammalian *diversity:* the Orders Carnivora (carnivores), Cetacea (whales), Artiodactyla (deer, antelope, cattle, etc.), Perissodactyla (horses, camels, rhinos, etc.), Rodentia (rodents), Lagomorpha (rabbits), Chiroptera (bats), and so on, all date back to the Eocene. The earlier Paleocene mammals largely died off, and only a few of the modern mammalian groups date back to the Mesozoic—groups, interestingly enough, such as our own order, Primates. That's right; we belong to one of the earliest, most primitive mammalian orders.

But it is, as I have already remarked, human evolution that most clearly disturbs creationists, and it is our own story that serves as perhaps the most dramatic and relevant of all examples of evolution. Human evolution perfectly fits the picture of ecosystem stability and disturbance, and the extinction and rapid evolution of species. We are like all other living creatures: until the invention of agriculture, like all other species, local populations of *Homo sapiens* and species ancestral to us had niches in the local ecosystems—initially in Africa, and later around the world. So it should come as no surprise that patterns in human evolution match perfectly the sorts of patterns seen in Paleozoic trilobites, Mesozoic dinosaurs, and early Tertiary mammals.

Spurred on by the intense interest in human evolution, paleoanthropologists have by now amassed a rich fossil record that is especially dense and illuminating for the past 3.5 million years or so. I can hardly do justice to the full story here;[10] rather, I will focus on the basic story of increase in brain size in human evolution, which along with increase in overall body size and the (largely earlier) adoption of an upright bipedal posture, is perhaps the most arresting feature of *progressive* evolution in hominid evolution. There simply is no doubt that brain size increases as we travel up through the rock record of human evolution in Africa. The tale of the skulls is creationism's worst nightmare.[11] And just like all other evolutionary patterns encountered in this chapter, the story is not one of gradual progressive change, but of episodes of evolutionary change following environmental disruption and, at least in one instance, extinction of earlier species.

I'll focus on just three critical phases to illustrate the point. Paleontologist Elisabeth S. Vrba of Yale University has noted the very same sort of bio-mere-like patterns in Africa seen by Palmer, Brett and Baird, and so many other paleontologists elsewhere throughout the past half billion years of life's evolutionary history; her term for the pattern is "turnover pulse." In a nutshell, Vrba has found that there was a dramatic worldwide drop in mean temperature of some 10 to 15°C that took place between 2.8 and 2.5 million years ago (no evidence of actual glacial ice in the northern continents of Eurasia and North America has so far shown up for this pre–Ice Age global cooling event). The effect of this cooling on the ecosystems of eastern Africa was a while in coming, but just about 2.5 million years ago, there was an abrupt change, from moist woodlands to a much drier, open grassland habitat.[12] Though some species were able to survive in wet woodlands elsewhere, and others, such as antelope that were already adapted to open grasslands, moved in (while still others, such as impalas, simply survived because they are adapted to both sorts of environments), the cooling effect wrought such drastic ecological change that many animal species simply became extinct. And because ecological disruption can trigger extinction as well as evolution (through speciation; see Chapter 4 for more on this), many new species evolved quickly as the new, open grassland ecosystem emerged.

Now, the dominant hominid species—in fact, the *only* hominid species we know about from the interval 3.0–2.5 million years ago—was *Australopithecus africanus* (literally "southern ape of Africa"). (So far, this species is known mainly from southern African deposits, where evidence of the same ecological events first worked out for eastern Africa has now been found.) *Australopithecus africanus* stood upright; males were no more than 4 feet in height and weighed no more than 100 pounds. Their pelves are distinctly more humanlike than apelike—part of the evolutionary change associated with the assumption of upright posture and bipedal walking. Their teeth are also distinctly more humanlike than apelike. And though some paleoanthropologists have claimed to see human features in the folds of the brains (endocasts) of this species, the actual brain size is firmly within chimpanzee range: about 400–450 milliliters in volume. For comparison, the average brain size of a modern human is approximately 1,350 milliliters.

*Australopithecus africanus* disappeared from the fossil record of southern Africa about 2.5 million years ago,[13] a presumed victim of extinction. Yet, just at 2.5 million years ago, we find (in both eastern and southern Africa) the remains of not one, but two, new lineages of fossil hominids. And we also find the first stone tools! Dramatically—if not really surprisingly—patterns of human extinction and evolution fit in *exactly* with the total picture of ecosystem disruption, extinction of many species, and the evolution of new, savanna-adapted species in their stead.

One of these two new lineages was the "robust" australopithecines, a fascinating lot indeed. The males had large sagittal crests for the insertion of massive jaw muscles needed to work their impressive lower jaws stocked with enormous molars for grinding their plant food. Though to the untrained eye the skulls of the several different robust australopithecines look somewhat gorilla-like, their brain size was about the same as that of *Australopithecus africanus*, and unlike any ape, they retained the structure of the pelvis and long bones inherited from their ancestors; in other words, robust australopithecines were true upright and walking hominids.

But they were definitely not our direct ancestors. Rather, they were an evolutionary side branch—ecological specialists obliged by their dental adaptations to eat tubers, nuts, fruits, and other vegetable materials. And, like other ecological specialists, the robust australopithecine lineage shows very rapid rates of evolution—with at least five different species occurring in Africa in a 1-million-year period (2.5–1.5 million years ago). The flip side of the rapid evolutionary rates of ecological specialists is their characteristically very high rates of extinction: none of the species of robust australopithecines so far discovered lasted very long.

The other lineage featured a more gracile anatomy: thinner bones, no massive crests on the head, smaller teeth—in other words, a skeletal anatomy far more like our own than like that of the robust australopithecines. The major difference between these newly evolved gracile hominids and their *Australopithecus africanus* predecessors was a significant increase in the size of the brain—to the vicinity of 700–750 milliliters. For this and other reasons, paleoanthropologists regard these as the earliest members of our own genus—the genus *Homo*. The earliest species, first occurring right at 2.5 million years ago in eastern Africa, was *Homo habilis*.[14]

Equally dramatic as the skulls themselves is the presence—again right at 2.5 million years ago—of the earliest definite signs of material culture. Here, at the very beginnings of the archeological record, we find the first stone tools—first discovered by the Leakeys in Tanzania's Olduvai Gorge.

Crude as they were, these ancient implements are thrilling, for they mark the first physical evidence that we can point to showing the departure, among our ancestors, from a purely biologically based to a learned behavioral (i.e., cultural) base for wresting a living from the surrounding local ecosystems in which they lived.

There is no question that the evolutionary events occurring just before and after 2.5 million years ago in Africa—events that were triggered by a global cold snap that disrupted Africa's ecosystems—had an enormous impact on human evolution, as well as on the evolution of many lineages of antelopes, pigs, birds, and so on. Without that "turnover pulse" (as Vrba refers to it), we simply would not be here today contemplating our own "origins," as the creationists like to put it.

But that's not all: the first Ice Age glaciation event—when huge sheets up to half a mile thick crept down from the northern polar regions to cover much of Eurasia and North America—also had its evolutionary effects, including on human evolution in Africa. For at just about this time, we find the earliest remains of what Ian Tattersall calls *Homo ergaster* (see Tattersall and Schwartz, 2000)—early African forerunner to the justly famous *Homo erectus,* which was first discovered in Java by the Dutch physician Eugene Dubois (1858–1940) near the end of the nineteenth century.

The thing about *Homo ergaster* is that, already by 1.5 million years ago, our ancestors had come to look very much like us. The best-preserved specimen discovered so far is the nearly complete skeleton of the so-called Nariokatone boy—thought to have been around 11 years of age at time of death. At full maturity, paleoanthropologists estimate he would have attained a full height of 5 feet 11 inches; in other words, he would have been as tall as the average present-day American male. As Ian Tattersall has put it, "from the neck down" *Homo ergaster* and *Homo erectus* were essentially modern in their skeletal anatomy. But not so their skulls—which contained a brain that, at about 1,000 milliliters, on the one hand was significantly larger than that of *Homo habilis,* but on the other hand still fell far short of the 1,350 milliliters that is the average volume of modern members of our own species, *Homo sapiens.*

In other words, as we move up in time through the African fossil record, we find species in our lineage with progressively larger brain sizes: from the 450 milliliters of the early australopithecines, through the 750 milliliters of the earliest members of the genus *Homo,* through later species *(Homo ergaster)* with brains of about 1,000 milliliters, finally culminating in the appearance of our own species, *Homo sapiens,* also in Africa, which evolved perhaps as much as 150,000–200,000 years ago, the oldest fossils occurring at about 120,000 years ago.[15]

So there it is: creationism's worst nightmare in a nutshell. There is a wonderful gradation of fossil human remains spanning an interval of the last 3 million years in Africa—so-called missing links occurring all the way as we go from something rather apelike (though human in its ability to walk upright) right on up through the appearance of modern human beings. We now see that successful early human species—like absolutely all other kinds of successful species that have been on Earth over the past 3.5 billion years—remained anatomically stable as long as their ecosystems remained stable. Physical environmental change, destabilizing those ecosystems, sent ancestral hominid species to extinction and spurred the evolution of new species just as it did in so many other plant and animal species of the ancient African landscape.

The evidence is just too good: ironically, the fossil record of human evolution is one of the very best, most complete, and ironclad documented examples of evolutionary history that we have assembled in the 200 years or so of active paleontological research. And, as we shall see (in Chapter 6), creationist attempts to punch holes in this incredibly well documented history of human evolution are simply puerile and downright wimpy.

The examples of evolutionary history presented in this chapter illustrate typical large-scale patterns in the history of life. Each shows us that evolutionary novelties—the appearance of the truly new—come at some point later in time than the appearance of the simpler, less complex anatomical structures. There is sense to the history of life—a sense made evident by the simple postulate that life has evolved. And that is the triumph of evolution.

# CHAPTER 4

# What Drives Evolution?
## The Evolution of Evolutionary Theory

If life has evolved, and if some kinds of evolutionary change can happen sufficiently rapidly to be observed during one person's lifetime, we should be able to go to nature to observe the change, to measure its tempo, and to test ideas about why and how it happens. And we should be able to simulate some kinds of evolutionary change experimentally on laboratory organisms, and even use mathematics and computers to run evolutionary experiments on an even more abstract level. This is the second meaning of the expression "evolutionary theory"—i.e., *how* does life evolve?—and the focus of discussion in this chapter.

That breeders can radically alter the anatomical and behavioral properties of animals and plants,[1] and geneticists can experimentally alter their fruit flies and guinea pigs, shows that small-scale evolutionary change can and does take place in brief intervals of time—obviously a circumstance consistent with the basic notion of evolution. Creationists often concede that such small-scale changes are possible, but they insist that these changes are not the stuff of large-scale evolutionary change between "basic kinds." But creationists are whistling in the dark.

The purpose of this chapter is to show precisely how evolutionary processes taking place in relatively small scales of space and time connect to larger-scale entities, processes, and events to produce the entire history of life—from the smallest incremental evolutionary change to the vast spectrum running from the simplest bacteria on up through the complex fungi, plants, and animals—from, in other words, the small-scale changes of so-called microevolution on up through the larger-scale changes often referred to as macroevolution. For, as we have already seen, this tremendously diverse array of life, spanning at least 3.5 billion years of Earth history, is all connected by a pattern of nested sets of genetic and anatomical similarity that can rationally be explained only as the simple outcome of a natural shared descent with modification. The only alternative is the decidedly vague and inherently untestable (thus inherently unscientific) claim that it simply suited a supernatural Creator to fashion life in this way.

Yet I have another motive in writing this chapter as well. I want to explore further the very nature of the scientific enterprise. It is no secret that scientists often disagree—sometimes over rather fundamental issues of their discipline.[2] So in this chapter I adopt a historical approach to reveal the growth of understanding of how the evolutionary process works, based on a prodigious growth of knowledge of all things biological (and geological) since the mid-nineteenth-century days when Darwin first convinced the thinking world that life has indeed evolved. But the road to enhanced knowledge and understanding is often bumpy, and I will reveal the arguments—and their resolutions—as I proceed.

It is also no secret that evolutionary biology still has its share of healthy disagreement. This is somewhat less true at the Millennium than it was, say, in the 1970s and early 1980s, since much has been accomplished since then. But there is still quite a lot to be resolved.[3] In particular, I will trace the growth of understanding of the evolutionary process with two goals in mind: first, to see how notions of *discontinuity* (so important to creationists, who insist there are no connections between "created kinds") have been incorporated into evolutionary thinking since the initial emphasis (going right back to Darwin) on *continuity*—among organisms within species, between species, and indeed in all of life. What discontinuities interrupt the continuous evolutionary spectrum from bacteria to redwood trees? What causes these discontinuities?

I also want to explore how we arrived at what I see as the greatest remaining source of conflict within evolutionary biology: the answer to the question, What actually drives the evolutionary process? As we shall see, all evolutionary biologists readily acknowledge that natural selection is the mechanism underlying adaptive change in evolution, and therefore the main ingredient of the evolutionary process that has produced the amazingly diverse array of life on Earth. And evolutionary biologists also agree that the evolutionary process is complex, involving ecological and physical environmental factors in addition to the genetic components that underlie the origin and maintenance of genetic variation, as well as selection for stasis and adaptive change.

Yet evolution is a phenomenon so rich and varied in its scope that biologists of many different disciplines come to look at the process from a variety of points of view. And therein lie some of the most exciting aspects of the study of evolution, and the source of many of the conflicts within the field. For example, the geneticist focuses, perforce, on short-term changes in the genetic composition of localized populations, while the paleontologist looks at a far broader sweep of evolutionary history: the fates of entire species and of even larger-scale groups. It is as impossible for me, a paleontologist, to study the generation-by-generation changes wrought by natural selection in my Devonian trilobites as it is for a geneticist to discover the patterns of regional ecosystemic disruption, followed by extinction and evolution of entire species. The job, clearly, is to integrate these multiscalar phenomena, and we're still not entirely there. But we've made a lot of progress—not just from Darwin's day, but even in the past twenty years.

Thus here is my millennial characterization of the grand dichotomy within evolutionary biology—stated in exaggerated form (with acknowledgment that many evolutionary biologists are already comfortable with some middle ground): On the one hand, some prominent evolutionary theorists tend to focus almost solely on natural selection. They see—correctly—that natural selection is the central process in evolution. But they tend to stop right there. The most extreme form of this position lies in the work of Oxford University biologist Richard Dawkins, author of the notion of the "selfish gene." Dawkins thinks that genes are in a constant competitive struggle among themselves to see their faithful copies represented in the next and succeeding generations. Evolution simply flows out of this competitive struggle.

Dawkins's characterization of natural selection is an extreme version of Darwin's original formulation in *On the Origin of Species* (1859):

> As many more individuals of each species are born than can possibly survive; and as, consequently, there is a frequently occurring Struggle for Existence, it follows that any being, if it vary however slightly in any manner profitable to itself, under the complex and sometimes varying conditions of life, will have a better chance of surviving, and thus be *naturally selected.* (p. 5)

In other words, Darwin saw that organisms compete for resources (mostly food—his "Struggle for Existence"), and that competition has implications for who reproduces. The genetic recipes for competitive success for organisms in the real world will tend to be passed along differentially, and when conditions change, other variant forms instead will be "selected."

However the Dawkins version of natural selection differs from Darwin's,[4] the main point here is that any evolutionary theory that focuses almost exclusively on natural selection—any evolutionary theory that does not specify the conditions that in effect turn natural selection on and off—is by definition an incomplete evolutionary theory. In a nutshell, Dawkins says that (1) natural selection amounts to competition among genes for representation in succeeding generations, and (2) since natural selection is the motor of change, the entire history of life flows from this fundamental first principle.

Thus Dawkins sees the primary impetus for evolutionary change as arising from the genes themselves. That's one of the two extreme answers to the question, What drives evolution? Here's the other: nothing substantial happens in terms of accruing adaptive evolutionary change *unless and until physical events upset the ecological applecart, leading to patterns of extinction and evolution of species.* This, it is obvious, is the position I tend to favor strongly—obvious in view of the condensed account of the 3.5-billion-year evolution of life that I gave in Chapter 3.

To my mind, the primary control over the evolutionary process lies in the events occurring in the physical world. I see a continuous spectrum from local ecosystem disturbance—with little or no evolutionary response—on up through the gross patterns of extinction and evolution occasioned by the five most severe episodes of mass extinction that have struck the planet in the last half billion years. I have deliberately stated these two starkly different views of the underlying causes of evolution in extreme terms. I now hasten to add that there is much merit in both: both have direct and very powerful relevance and application to the specific domains from which they arose, and which they endeavor to explain. The Dawkins gene-centered version is especially powerful on the small-scale level of generation-by-generation change within populations, and it is, in addition, a vital component of the underlying theoretical structure of the subdiscipline of sociobiology. On the other hand, it is useless as a general evolutionary theory covering the large-scale events in the history of life.

Then again, though, emphasis on large-scale climatic and other physical events yields an evolutionary perspective that does take the large-scale biological historical events into account, but is weak in terms of explaining the internal dynamics of selection (and other genetic phenomena) on smaller scales. Clearly, we need a general evolutionary theory that brings both elements into play. The remainder of this chapter, through its historical account, will seek to build to such a pluralistic view—with the recognition that there is still much work to be done. This is, after all, science, and science is a learning process. We have come a very long way, but there is still much more to be learned.

### The Evolution of Evolutionary Theory

Darwin relied on a barrage of patterns to convince his readers that life had evolved. These included the grand pattern of nested resemblance absolutely linking up all forms of life, as well as patterns in the geographic and geological distribution of organisms. But it is generally supposed that Darwin succeeded where others before him had failed in establishing the credibility of the very idea of evolution because he supplied, as well, a plausible mechanism to explain *how* life evolves.[5]

We have already encountered Darwin's original exposition of the principle of natural selection: in a nutshell (once again), because more organisms are born each generation than can possibly survive and reproduce,[6] and because organisms vary within local populations (and those variations are heritable), on average, those organisms best suited to coping with life's exigencies are the most likely to survive and reproduce. It follows as a logical necessity that the heritable recipes for success (what we now understand to be the genetic information underlying more successful adaptations) will tend to be passed along differentially to the next succeeding generation.

It is important to realize that Darwin's theory—pangensis—about how organisms tend to resemble their parents (in stark contrast to his views on how life evolves) was later shown to be false. In Darwin's time, there was no science of genetics, indeed no concept of the gene. Darwin's views on inheritance belong in the same realm of outmoded scientific concepts as the idea of a flat Earth, or the notion that the sun revolves around our planet. His theory of inheritance envisioned various organs of the body each contributing "gemmules," little forerunners of each cell type in the body, which congregate in eggs and sperm and develop into parental replicas in the offspring. We now know that the same information that serves as a recipe for development in the fertilized egg is found in each[7] of the billions of cells in any vertebrate organism's body. Darwin knew nothing of this, but as it turned out, his ignorance was sublimely irrelevant to the problem he was really interested in tackling: evolution.

In other words, the knowledge that organisms show variations, that those variations are heritable, and that population sizes are inherently limited was sufficient for Darwin (and his contemporary, the naturalist Alfred Russell Wallace, 1823–1913) to formulate the notion of natural selection. It doesn't matter, it turns out, how the mechanisms of heredity actually work to understand how natural selection works. This point was not fully grasped by biologists: Many early geneticists, at the dawn of the twentieth century, thought that their discoveries of the fundamental principles of genetics somehow cast doubt, or rendered obsolete, the concept of natural selection. It took several decades of experimentation and theoretical (including mathematical) analysis to show not only that there was no conflict inherent between the emerging results of the new science of genetics and the older Darwinian notion of natural selection, but that the two operate in different domains. The principles of inheritance work within single organisms—two organisms, in the case of sexual reproduction. In contrast, natural selection involves differential reproductive success among large numbers of genetically varying organisms within a reproductive population.

Darwin touched off a storm when he published *On the Origin*. Today's creationism is, in some sense, merely an echo of that distant thunder. But many of the biologists who were immediately convinced *that* life had evolved took sharp issue with Darwin's notions on *how* life had evolved. Natural selection seemed to some a bit too brutish, too materialistic. And still other biologists were attracted to alternative theories on how life might have evolved. Foremost among the competing theories was Lamarckism.

Jean-Baptiste Lamarck (1744–1829) was a French zoologist active a half century prior to the appearance of Darwin's book. A great zoologist convinced of the interconnectedness of all life, Lamarck is generally remembered today (largely in scorn) mostly for one tiny fragment of his intellectual edifice: he thought that evolutionary change occurred when an organism developed a new capability, or underwent a slight modification, during the course of its lifetime, and subsequently passed the newly acquired trait along to its offspring. The classic conflicting scenarios of "how the giraffe got its long neck" (with apologies to Rudyard Kipling) demonstrate the differences between the Lamarckian notion of the inheritance of acquired characteristics and Darwinian natural selection.

The Lamarckian version sees individual proto-giraffes stretching their necks, craning for leaves in tree canopies, and passing on their slightly distended necks to the next generation. In contrast, the Darwinian view invokes variation in neck length within a population of proto-giraffes and imagines a situation in which competition for leaves high up in tree canopies selectively favors those with the relatively longer necks. These longer-necked giraffes tend to survive and leave more offspring, and little by little the average neck size of the proto-giraffe lineage increases.

It is a testimony both to Victorian biology's ignorance of genetics and to the persuasiveness of Lamarckian-inclined biologists that Darwin allowed Lamarckian notions to creep into later editions of *On the Origin*, though he steadfastly maintained his preference for natural selection. But in the

1880s, just after Darwin's death, the German biologist August Weismann (1834–1914) propounded what we now call the Weismann doctrine: that germ cells (eggs and sperm) create the physical body, and that the process is a one-way street. Nothing that happens to the body turns around and affects the germ cells. Characteristics developed during an organism's lifetime simply cannot be transmitted to offspring. Weismann, in other words, claimed to have refuted Lamarckian notions of evolution.[8]

In any case, Weismann had convincingly removed Lamarckian notions from biology just when genetics was in its birth throes in the early years of the twentieth century, when three biologists nearly simultaneously discovered the works of the Austrian monk Gregor Mendel (1822–1884). Mendel's famous experiments with peas implied that the factors that underlie the inheritance of features—say, blue or brown eyes, or wrinkled or smooth pea skins—are particulate, since there is a marked degree of independence and shuffling of these characteristics that seem to follow simple laws when careful breeding experiments are performed. Within a decade, the science of genetics was up and running, and important discoveries, such as that genes lie in linear arrays on chromosomes in the nuclei of all but the simplest cells, had already been made.

All the rapid advances made in the heady days of early research in genetics seemed to challenge Darwin's idea of natural selection. In fact, most biologists concentrated on such fields as genetics and the equally rapidly advancing field of physiology, spurning evolutionary biology altogether as old-fashioned. And the biologists who were left to ponder the mysteries of evolution professionally in the early decades of the twentieth century were troubled: how could Darwin's scheme of natural selection, which he saw as working on a smoothly gradational field of variation, be reconciled with the idea that genes are separate particles, each producing its own characteristic? And what of the Dutch botanist Hugo de Vries's (1848–1935) notion of mutations, in which large-scale changes in the flower of the evening primrose would suddenly appear?

"Mutations" came to be the official term for what the breeders had for years more informally termed "sports'"—the unexpected, quirky appearance of new features not inherited from parents or, for that matter, grandparents.[9]

Mutations are the source of all true novelty, it seemed, but they also seemed to be large-scale effects, not the minor new forms of variation required by the classical Darwinian view.

In the context of the new genetics, natural selection seemed to be in trouble. Many conflicting versions of evolutionary theory sprang up in the first two decades of the twentieth century, some of which not only violated Darwinian principles but also were inconsistent with the new genetics.[10] By the 1920s, biology had drifted rather far from the Darwinian fold, though evolutionists were quick to rally around the flag when the call went out for expert witnesses (who actually never ended up on the stand) for the Scopes trial.

Thus, just when the Scopes case came bursting onto the national stage, evolution was a relatively minor area of biological research taken as a whole and was, as well, fraught with difficulties. Some paleontologists in the 1920s still clung to Lamarckian notions (despite Weismann's valiant efforts to debunk Lamarckism), while others adopted vitalism, a mystical notion that saw evolution as an inner-directed drive to perfection. But it was the apparent conflict between genetics and Darwin's notion of natural selection that was the real stumbling block.

It was the work of three mathematically gifted geneticists—the American Sewall Wright (1899–1988) and the Englishmen J. B. S. Haldane (1892–1964) and Sir Ronald Fisher (1890–1962)—that resolved the apparent incongruities between the mechanisms of inheritance and the principle of natural selection. By the late 1920s, it had become clear that many mutations have relatively minor effects, that some genes play a role in forming more than one characteristic, and, most importantly, that most characters are formed by the action of more than one gene. The old black-white, brown-blue, wrinkled-smooth, either-or dualities in the theory of inheritance had given way to the view that a spectrum of variation (in height, say, or number of head hairs) could nonetheless be under genetic control. All of a sudden, there were no longer any formal objections, from the genetics quarter, to the notion of natural selection working to preserve the most beneficial of a spectrum of variation in each generation.

All this came from genetics, despite complete ignorance of the biochemical "anatomy" of genes—the structure of DNA. The biochemical basis of inheritance was still a black box, but at least what geneticists had learned circa 1930 no longer seemed to render natural selection an impossibility. The way was thus cleared for a rapprochement between genetics and evolutionary theory. Wright, Fisher, and Haldane made important contributions—essentially founding the field of population genetics, in which the effects of selection, mutation rate, and random genetic drift[11] could be studied by mathematical formulas and the analysis of experimental results.

But it was the Russian-born geneticist Theodosius Dobzhansky (1900–1975), an immigrant to New York in the 1920s, who really put it all together. Trained as a naturalist (actually, as a beetle specialist), Dobzhansky took up the study of fruit flies, which had already yielded such stunning experimental results in Thomas Hunt Morgan's (1866–1945) laboratory at Columbia University in the first decade of the twentieth century. In the 1930s, Dobzhansky began a long series of studies of natural populations in the wild of fruit fly genetics—a series that continued right up to his death in 1975.

In 1937, Dobzhansky published his first major book, *Genetics and the Origin of Species*, a title with a deliberate reference to Darwin's greatest book, published so long before. And the bridge Dobzhansky built involved far more than a literary allusion. In the book, Dobzhansky effectively argued for the central role played by natural selection in changing the genetic composition of populations of organisms in nature.

Darwin emerged completely vindicated: natural populations seem to possess ample amounts of genetic variation. But this variation, as Darwin himself had noted, seems highly organized: local populations within a species differ from place to place, reflecting adaptation to slightly different environments. The average length of a sparrow's wing, for example, is longer in warmer climes than in populations situated high up the slopes of mountain ranges. The physiological explanation is simple: shorter wings radiate less body heat, a distinct advantage in colder climates. Natural selection, then, preserves the variants within species that are best suited to the precise ecological conditions of each local habitat over the entire range of a species.

Dobzhansky's book accomplished a lot more than finishing the fusion between genetics on the one hand, and the original Darwinian vision of adaptation through natural selection on the other. In his first chapter, Dobzhansky pointed out that, if that were all to the story, the world of living creatures would look a lot different to us than it does. Looking around us, we see populations of different species of birds, mammals, trees, and shrubs. Thinking of the denizens of my own backyard for a moment: The gray squirrels all look pretty much alike (despite some underlying genetic variation), and they look very different from their close relatives, the eastern chipmunks. Likewise individual specimens of the several species of oaks and maples all look a lot alike—and quite different from other, closely related tree species.

In other words, Dobzhansky looked around and saw *discreteness* in the living world. He saw gaps between species—gaps that simply wouldn't be there if natural selection, working on a spectrum of variation within one enormous species population, were all there is to the evolutionary process. Natural selection, Dobzhansky knew, would be expected to produce a continuous array of variation, since populations—and entire species—are adapted to the specific environments of each and every locale.

Dobzhansky, moreover, had the first insights on how the hierarchical structure of living systems is related to the evolutionary process. At the level of the genetics of individual organisms, he realized that genes are particulate and discrete, as are their variant versions (alleles). (What we have learned about the structure and composition of DNA and RNA in the years since Dobzhansky wrote his book only bears this point out further.) At the second level, however—the population level—things are completely different: here we enter the realm of relative gene frequencies, in which selection (and genetic drift) shift frequencies of genes in a continuous, gradational manner. Here lies the real reason why Darwin could formulate accurately his theory of natural selection without knowing the first thing about genetics: the two processes—inheritance and changing gene frequencies through natural selection and genetic drift—take place at completely different levels of biological organization.

There is a third, even higher level, in which populations of organisms are aggregated into larger arrays that we call species. Dobzhansky (and his contemporary, American Museum of Natural History biologist Ernst Mayr,

1904– ) conceived of species essentially as the largest collections of organisms that are capable of mating with each other. And these reproductive communities, these species, are for the most part discrete entities. At this higher level, once more we confront gaps and discreteness.

Darwin of course saw the same gaps, but because his primary goal was to convince the world that life had evolved, he quite naturally stressed the evidence for continuity, downplaying discontinuity as much as an honest man like he possibly could. After all, he was trying to overthrow a worldview that was beautifully summarized in two of my favorite quotes from pre-evolutionary, pre-Darwinian days: (1) "Species tot sunt quot ab initio faciebat Infinitum Ens (meaning "There are as many species as originally created by the Infinite Being"), by Carolus Linnaeus, the great botanist and founder of the modern scheme of classification, writing in *Systema Naturae* (1758); and (2) "Species have a real existence in nature, and a transition from one to another does not exist" (a statement from British philosopher William Whewell, writing in *History of the Inductive Sciences*, 1837).

Both statements acknowledge the reality and discreteness of species and maintain that species are not interconnected by a natural process of evolution. Darwin obviously felt that, to establish that species are connected in an evolutionary sense, he had to deny, in a very real sense, their discreteness. That there are gaps between species was plain enough to Darwin, but he chose to explain the gaps purely by invoking the extinction of intermediate forms.

Dobzhansky thought otherwise. He thought that gaps between species are the direct result of the evolutionary process itself. In answer to the question of why the living world is cut up into an array of (some 10 million) more or less discrete species, Dobzhansky maintained that partitioning of pools of genetic information would help focus a species' adaptations to the environment: one large species, spread out over a variety of environments, would contain the genetic information pertinent to each localized environment. And that information would spread (in a process called gene flow). To focus more narrowly on a subset of environments, it would be helpful, Dobzhansky thought, to interrupt that gene flow, to partition that vast array of genetic information—in short, to make two species where there once had been but one.

Dobzhansky also spoke of a natural trade-off inherent in the genetics of natural biological systems (i.e., populations and species): in general, it is best not to focus adaptations (and underlying genetic information) too narrowly, lest the environment change and extinction loom. On the other hand, a continuous array of genetic information would not support adaptive focus on any subset of environmental conditions.

That is why Dobzhansky's book is best remembered for its analysis of the origin of new species. Ironically, Darwin never did discuss the origin of species in his book of that name. For reasons just discussed, to Darwin, the "origin of species" was essentially synonymous with the term "evolution" itself. The simple accumulation of anatomical change over thousands and millions of years was what Darwin had in mind when he thought of new species arising from old. Occasionally, he realized, species would somehow become fragmented, such that what was once a single species would leave two (or even more) species, each newly embarked along slowly divergent, separate pathways. But to Darwin, such divergence was little more than a special case of natural selection molding far-flung populations along somewhat different lines, until their differences appeared so great that we would be obliged to call them different species.

Dobzhansky and other biologists in the 1930s recognized that species are something more than mere collections of organisms similar in their anatomical and behavioral properties. They are also breeding communities; in fact, it is their interbreeding, and consequent sharing of genetic information, that makes their component organisms so similar in the first place. Years of observation in the field and experience in the lab and breeder's pen showed that most species cannot be successfully mated even with their closest relatives. More to the point, even when such hybridization is technically possible—when the species are found together in the wild—hybrids are nonetheless rare, or altogether absent; i.e., even if they can interbreed, they don't. Lions and tigers in India (yes, there still are a few lions in India) have been in close contact in some places within human memory, yet, insofar as I am aware, the "tiglons" produced in some zoos have never been seen in the wild.

How do new reproductive groups—new species—originate? Dobzhansky's thoughtful discussion of the origin of what he called isolating mechanisms has settled the issue. The behavioral and anatomical differences between species are what keep them from interbreeding. These differences usually evolve when a species is so far-flung that some of its populations become isolated from the main group. In other words, normal adaptive differentiation (as with wing length in sparrows), plus prolonged geographic isolation (when no mating between the two divisions of a species takes place), starts the ball rolling. If divergence has proceeded far enough, when two groups that used to be members of the same species once again come into contact, genetically based anatomical, physiological, and behavioral traits have become so differentiated that the groups no longer interbreed.

Ernst Mayr, an ornithologist working at the American Museum of Natural History in New York, published *Systematics and the Origin of Species* (1942) a few years after Dobzhansky's book appeared. Mayr knew that naturalists even before Darwin were aware of a general pattern in nature in which a species' closest relative typically lives in an adjacent region—not in the same area. Victorian naturalists called such species vicars (in the sense that they either replaced or represented one another geographically) or germinate ("twin") species. In a sense, such species are just slightly different versions of the same thing. The scarlet tanager of eastern North America, for example, is "replaced" in the West by the equally striking, if differently colored, western tanager.

Darwin used the patterns of geographic variation and vicariant species as part of his argument that life must have evolved. In contrast, Dobzhansky, Mayr, and other evolutionists of the 1930s and 1940s emphasized the importance (in fact, the utter necessity) of geographic isolation for disrupting an ancestral species to produce two species where once there had been but one. To these later workers, evolution is descent with modification all right, but it is produced by a one-two combination of adaptive change plus occasional episodes of disruption, creating the great array of different species we see all around us today. Thus by the early 1940s, we finally had a theory of how natural selection effects genetic change, and how accumulated genetic change plus geographic isolation yields reproductive isolation—a new species derived from an older, ancestral species. Biologists had finally added the all-important aspect of species-level discontinuity to the spectrum of continuous variation generated by natural selection.

But what of the large-scale changes in evolution? The work of Fisher, Haldane, Wright, Dobzhansky, and Mayr (and, of course, their many colleagues) spoke directly to the evolution of the sorts of differences between, say, lions and tigers. The conclusion that the kind of adaptive change within species underlies the differences we see between closely related species has long since seemed so voluminous and incontestable that even today's creationists accept microevolution. Patterns of evolutionary change within species seem no different in principle—just milder in degree—from the sorts of changes we see between closely related species. All evolutionary changes are produced by natural selection working each generation on the variation presented to it. Could this simple process also account for all the changes among the larger groups of animals and plants—between, say, reptiles and mammals?

Let us return just for a moment to the grand pattern of nested sets of resemblance linking up absolutely all species on Earth. Closely related species are classified in the same genus, and related genera are put in the same family. Families are lumped together in the same order, and related orders are grouped into the same class. This is the Linnaean hierarchy that was devised, long before Darwin's day, to encompass the nested pattern of similarities that define groups of organisms—a pattern naturalists had observed in the organic world since Aristotle. But, we must ask, what exactly are these genera, families, orders, and so on? It was clear to Darwin, and it should be obvious to all today, that they are simply ever larger categories used to give names to *ever larger clusters of related species.* That's all these clusters, these higher taxa, really are: simply clusters of related species.

Thus, in principle the evolution of a family should be no different in its basic nature, and should involve no different processes, from the evolution of a genus, since a family is nothing more than a collection of related genera. And genera are just collections of related species. The triumph of evolutionary biology in the 1930s and 1940s was the conclusion that the same principles of adaptive divergence just described—primarily the processes of mutation and natural selection—going on within species, accumulate to produce the differences we see between closely related species—i.e., within genera. Q.E.D.: *If adaptive modification within species explains the evolutionary differences between species within a genus, logically it must explain all the evolutionary change we see between families, orders, classes, phyla, and the kingdoms of life.*

This highly reasonable inference still demanded cogent exposition. In his *Tempo and Mode in Evolution* (a book begun in the late 1930s but not published until 1944), George Gaylord Simpson (1902–1984), a vertebrate paleontologist also at the American Museum of Natural History, attempted to show that all the major changes in life's evolutionary history could be understood as a by-product of these newly understood principles of genetic change. To undertake this task, Simpson had to confront that greatest and most persistent of paleontological bugbears: the notorious gaps of the fossil record.

Darwin and his evolution-minded successors in the paleontological ranks preferred to explain these gaps away: they blamed the incompleteness of the geological record of the events of Earth history. According to this explanation, the lack of abundant (creationists say *any*) examples of smoothly gradational change between ancestors and descendants in the fossil record merely bears witness to the gaps in the quality of the rock record. True enough, but not entirely so, said Simpson, to his everlasting credit.

Simpson thought the fossil record has a great deal to say about how evolution occurs—its pace and style, its "tempo and mode." After all, it is in the enormous expanse of geological time that the evolutionary game has actually been played. But to make a such a claim is also to assert that the fossil record is at least complete enough to be taken seriously. Thus the gaps had to be confronted. And since gaps there certainly are, they must at least in part be a product of the evolutionary process, if not merely the artifacts of a poor geological record.

It is the gaps in the fossil record that, perhaps more than any other facet of the natural world, are dearly beloved by creationists. As we shall see when we take up the creationist position, there are all sorts of gaps: absence of gradationally intermediate transitional forms between species, but also between larger groups—between, say, families of carnivores, or the orders of mammals. In fact, the higher up the Linnaean hierarchy you look, the fewer transitional forms there seem to be. For example, *Peripatus*, a lobe-legged, wormlike creature that haunts rotting logs in the Southern Hemisphere, appears intermediate in many respects between two of the major phyla on Earth today: the segmented worms and the arthropods. But few other phyla have such intermediates with other phyla, and when we scan the fossil record for them we find some, but basically little, help. Extinction has surely weeded out many of the intermediate species, but on the other hand, the fossil record is not exactly teeming with their remains.

Simpson knew this but preferred a view of evolution consistent with the emerging principles of genetic change over the alternative posed by German paleontologist Otto Schindewolf (1896–1971). Schindewolf interpreted the gaps in the fossil record as evidence of the sudden appearance of new groups of animals and plants. Not a creationist, Schindewolf believed all forms of life to be interrelated, but he felt that the fossil record implies a saltational mode of evolution—literally, sudden jumps from one basic type (called a Bauplan, or fundamental architectural design—conceptually if tangentially related to creationists' "basic kinds") to another. Simpson and his peers scoffed at such an idea, and rightly so, since little evidence emanating from genetics laboratories even remotely hinted at how such sudden leaps could occur. And the cardinal rule of science—that all ideas must be testable—held sway: the prevailing theory and evidence of genetics of the 1930s (and genetics hasn't changed all that much since) were against large-scale, sudden switches in the physical appearance of descendant organisms. Schindewolf's views were at odds with nearly all that was known of genetics in the 1930s. His saltational explanation of the gaps was impressive—but wrong, as far as Simpson was concerned.

Simpson thought that most of the fossil record amply supports Darwin's view—that species slowly but surely—*gradually*—change through time, such that new species arise from old by imperceptible gradations over time. There is plenty of evidence, he felt, to show that 90 percent of evolution involves the gradual transition from one species to the next through time. When there were gaps between closely related species and genera—in other words, when new species appear abruptly in the fossil record with no smoothly intergradational intermediates between them and their ancestors—Simpson was (like Darwin) content to blame the gaps on the vagaries of preservation inherent in the formation of the fossil record.

In contrast to his treatment of species, Simpson did acknowledge that the sudden appearance of new groups—those ranked rather high in the Linnaean hierarchy, and what the creationists call basic kinds—implies something *true* about how evolution works. If evolution were always a slow, steady change from species to species, Simpson pointed out, the transitions between major groups would typically take millions of years, and we should expect to find some fossil evidence of the transitional forms. Not finding them very often, he deduced, implies that evolution sometimes goes on rather quickly—in brief, intense spurts. The presence of some intermediates (such as *Archaeopteryx*, the proto-bird) falsifies Schindewolf's saltational notions. But the relative scarcity of such intermediates bespeaks a major mode of evolution producing truly rapid change—a mode Simpson called quantum evolution. In physics, a quantum is a sudden, definite shift in state, a jump from one state to another without intermediates, an either-or proposition. The fossil record mimics these sudden changes in state, as is shown, for instance, by the evolution of whales from terrestrial mammalian forebears.

Whales first appeared in the Eocene, some 55 million years ago. They are primitive—for whales. But the earliest specimens look like whales, and it is only their general mammalian features that tell us they must have sprung from another group of terrestrial mammals in the Paleocene.[12] Bats are another example: a perfect specimen from the Eocene of Wyoming is primitive—for bats—but anyone looking at it will see at once that it is a bat, and evidence for its derivation from a particular kind of insectivorous Paleocene mammal still hasn't turned up.

Simpson's specific theory about how these sudden shifts come about in the course of evolution is no longer accepted in all its details, mainly because Simpson himself effectively retracted the idea in his sequel, *The Major Features of Evolution*, published in 1953. The original idea of quantum evolution envisioned a small population rapidly losing the adaptations of its parental species, going through a shaky phase, and then luckily hitting on a new set of anatomical and behavioral features suitable to life in a new—radically new—ecological niche. Simpson supported his theoretical stance with the solid structure of genetics, but he later dropped the notions of "inadaptive" and "preadaptive" phases. Why he did so reveals the power of the very idea of natural selection.

The tide of thinking in genetics was simply running against notions of genetic change that didn't specifically include natural selection. Early in the 1930s, for instance, the geneticist Sewall Wright spoke about the fate of genetic information among different breeding populations (he called them demes) within a species and of the role that chance plays in determining the composition of the next generation. Wright was not wholly infatuated with natural selection as the only agent of evolutionary change. Although his concept of genetic drift as a random factor in evolution was accepted by his peers, it was a grudging acceptance. It is only recently that his shifting-balance theory, in which the differential success of such breeding populations is a major issue, has once again begun to be taken seriously.

Both notions pay but little heed to the narrow, restricted version of natural selection, which is differential reproductive success of individuals within breeding populations based on their relative abilities to cope with life's exigencies. But by and large, Dobzhansky's book, plus Mayr's and Simpson's and a host of shorter technical publications in the late 1930s and 1940s, came close to asserting that the only significant force underlying genetic, hence evolutionary, change is natural selection. By 1953, Simpson himself saw his original concept of quantum evolution as too far removed from the consensus. He modified quantum evolution to mean merely an extremely rapid phase of change, governed throughout by natural selection as a small population invaded a new habitat.

By the early 1950s, the gist of evolutionary theory said that genetic change is largely a function of natural selection working on a field of variation presented to it each generation. New features from time to time appear, ultimately brought about by mutation. Most mutations are harmful; some are neutral, or even beneficial. The neutral or beneficial mutations hang on and one day might prove to be a real advantage as the environment provides new challenges to the organisms. In any case, as the environment changes (as all environments eventually do), it does so generally slowly. Natural selection preserves the best of each generation, and their genes make up the succeeding generation. Through time, through enough generations, selection wreaks tremendous changes. And occasionally habitats divide and species fragment, following separate adaptive histories. Hence, species multiply. Given enough time (and remember that geologists tell us that the

Earth is fully 4.5 billion years old), all manner of change will accrue: species keep on giving rise to new species, and adaptive evolutionary change accumulates. The more species, the more adaptive change—and the greater the evolutionary divergence. Genera, families, orders, classes, and phyla—all owe their genesis (as it were) to the simple processes of mutation, natural selection, and speciation.

Thus, by 1959, the centennial year of the publication of Darwin's *On the Origin of Species*, the essence of Darwin's vision had been integrated with the science of genetics. The importance of the evolution of new species, producing a discontinuous array of adaptive diversity, had been added, though it was already being downplayed somewhat and was soon to be ignored almost completely by a major, gene-centered branch of evolutionary biology that developed in the 1960s and 1970s. Though most evolutionary biologists were congratulating themselves and each other that, not only had Darwin been vindicated but also in effect a complete evolutionary theory had already been achieved, post-1959 developments were soon to belie that conclusion.

## Evolutionary Biology at the Millennium

The molecular revolution that began with the famous discovery by Watson and Crick (and unheralded partners, such as Rosalind Franklin) of the structure of DNA has had a pronounced, if subtle, effect on evolutionary biology during the last forty years of the twentieth century. Indeed, in a sense history replayed itself, as some molecular geneticists thought that the new understanding of the molecular structure of the gene somehow alters the older formulations of population genetics. But, as we have seen especially with the work of Theodosius Dobzhansky—who so clearly saw that the genetic processes of inheritance take place at the level of the individual, whereas natural selection and genetic drift take place within populations of individual organisms—any new knowledge of the chemical structure of genes can only enrich, rather than supplant, the understanding of evolutionary processes that has accumulated since the days of Charles Darwin.

And molecular biology has indeed enriched evolutionary discourse. For a while, controversy erupted over ways in which molecular processes could themselves bias the transmission of forms of genetic information,[13] suggesting to some biologists that fundamental concepts of the evolutionary process would have to be modified in light of the new science of molecular genetics. Among the more subtle impacts of the molecular revolution in evolutionary biology, however, was the emergence of one of the main strands of evolutionary thought of the past forty years, which was mentioned at the outset of this chapter:[14] the approach to evolution epitomized by Richard Dawkins's concept of the selfish gene. This explicitly gene-centered movement began right after the Darwinian centennial, with the publication of two critical papers by geneticist William Hamilton (in 1964), and especially American biologist George C. Williams's influential book *Adaptation and Natural Selection* (1966). Williams's work attempted to inject greater rigor into evolutionary analysis and maintained that natural selection works strictly for the "good of the individual"—i.e., not for the good of the species, as evolutionary biologists had been wont to assume, rather uncritically, off and on ever since Darwin.

Perhaps the most important aspect of this work has been the development of the field of sociobiology. To Darwin the presence of altruism, in which organisms act to benefit others in their populations, is a bit of a conundrum: if, he reasoned, the continual struggle for existence constantly pits organisms against each other for survival (and ultimately for reproductive success), what advantage could there be in *helping* other organisms? Hamilton's twin papers in 1964 went far toward resolving this problem, which was emerging as a real difficulty, given the renewed emphasis on individual selection (as exemplified by Williams's book in 1966).[15] Hamilton showed by mathematical analysis that the more genes any two organisms share, the more it is to their own benefit (i.e., in terms of seeing their own genes make it to the next generation) to cooperate with their relatives.

•

Social systems, of course, rely on cooperation among their members. The entire field of sociobiology—spearheaded by Harvard evolutionary biologist E. O. Wilson—sprang from the realization that altruistic behavior is expected to be correlated with degrees of genetic relatedness. This construct works particularly well with social insects, which are characterized by peculiar genetic systems in which nestmates share even higher percentages of their genes than the 50 percent genetic similarity typical of siblings.

Sociobiology is only one conspicuous example of the focus of post-1959 evolutionary research. There have also been detailed examinations of natural selection in the wild, as well as numerous studies documenting the intense evolutionary interactions among different species in local ecosystems.[16] The major journal devoted to evolutionary biological research in the United States—*Evolution*—greatly expanded the number of pages published annually during the last half of the twentieth century. Modern evolutionary biology, focused on genes and natural selection, is alive and well, and it has fully incorporated molecular genetics into its fundamental structure.

Indeed, we now know that genetic, evolutionary change in natural populations is occurring constantly, and at least in some instances the genetic flux is so great that the problem is to reconcile it with one of the great patterns in the evolutionary history of life briefly encountered in Chapter 3: stasis. How can genetic change be so common and so rapid within populations, while species as a whole tend to remain fairly stable over millions of years? Then, too, though some geneticists have turned their attention to the speciation process to great effect, on the whole the focus on within-population genetic change has tended to downplay the significance of the discontinuities between species that had been added to the evolutionary discourse by Dobzhansky and Mayr just before World War II. But those gaps are real, and they are created by the evolutionary process. And both these problems—stasis and the creation of gaps between species by the very process of speciation—came together in the theoretical postulate known as punctuated equilibria, part of a second major stream of evolutionary research that developed in the 1960s. This field of research stresses the larger-scale patterns in the history of life, many of which were encountered in Chapter 3.

In the 1960s, I was one of a number of graduate students interested in evolutionary paleontological research. Most of us were enrolled in Columbia University's Geology Department, for though our primary interests were distinctly biological, invertebrate paleontology had traditionally been a field of geological investigation. And it was invertebrates that caught the eyes of most of us, since invertebrate fossils are found in great profusion in the Paleozoic, Mesozoic, and Cenozoic strata of the world; often, many thousands of specimens can be collected that span millions of years. And one thing that evolutionary theory had long since made abundantly clear is that evolution through natural selection works on variation within populations of organisms. We also understood that geographic variation is an important ingredient of the evolutionary process, and many marine formations crop out over truly vast areas on the world's continents.

So naturally we thought it would be far easier to study evolution if we had large samples of brachiopods or clams or snails—or, in my case, trilobites. Studying evolution would be a snap, we thought: all we needed to do was drive from outcrop to outcrop, plotting out patterns of within-population and geographic variation, tracing evolutionary history directly as it unfolded before our eyes as we sampled up through successfully younger layers of fossiliferous rock. Basically, the evolutionary pattern we expected to find was Darwin's originally predicted pattern: slow, steady gradual change within species as we looked up the geological column.

At first, I was frustrated. Collecting trilobites from New York west to Iowa, from southern Canada south through Virginia, sampling populations that spanned at least 6 million years, I could find very few differences among the hundreds of specimens I found. Despair began to creep in, as I thought my first major piece of paleontological research was turning out to be a failure. As I struggled to find some signs of gradual evolution within my lineage of Devonian trilobites, it finally became clear to me that stability was the norm, the rule in the history of my species lineages of trilobites.[17] All around me, others were finding great stability in their species lineages of fossils as well.

Yet evolution was nonetheless plainly occurring: new species of trilobites, distinguished by small but consistent differences between samples, occasionally did appear. When I placed my data on a series of maps representing five successive slices of time, a pattern of geography leaped out, instantly suggesting that my trilobites were evolving through the general process of geographic speciation, as described by Theodosius Dobzhansky and Ernst Mayr. I described these results in the journal *Evolution* in 1971, and a year later my fellow graduate student (since gone on to the Harvard faculty) Stephen Jay Gould and I developed a general model of evolution based on my trilobites, his snail examples, and what we took to be fundamental patterns of (1) stasis and (2) geographic speciation that we thought were demonstrably the rule rather than the exception in the evolution of species throughout the history of life (see Eldredge and Gould, 1972).

The gaps between species are especially important here, given creationist claims that gaps between species (and higher taxa) are sufficient to disprove the very notion of evolution. Though some of my samples of Devonian trilobites do preserve some transitional states between ancestral and descendant species, by and large what we paleontologists see is the persistence of stable species for millions of years, followed by the appearance of other, usually quite similar species. Sometimes there is a clear geographic component to the pattern, but very often we find these species occurring one on top of the other in the rock column. Sometimes the two (i.e., ancestral and descendant species) overlap in time, but very often they do not.

How could we explain these gaps—these quick changes from one set of anatomical features to another? Easily: Gould and I realized that standard geographic speciation theory as developed especially by Dobzhansky and Mayr was sufficient. Geographic isolation leading to reproductive isolation need not take long to occur: our estimate was from five thousand to fifty thousand years, and some geneticists immediately commented that speciation could go on a lot faster than that. Speciation can be a rapid process— and one *altogether too quick to show up in detail in the fossil record*. The layers of sediment that entomb the fossil record are simply too episodic, too inherently gappy themselves, to record the passage of every successive year—or decade—or century—of geological time. In other words, the naturally occurring gaps between closely related species in the modern fauna and flora, directly caused by the process of fissioning known as speciation, typically happens so quickly that rarely do we catch it in midstream when we scour the fossil record for insights on how evolution occurs.

But that's not all. When we published our paper on punctuated equilibria in 1972, although Gould and I claimed the pattern was very general—typical of species up and down the fossil record since time immemorial—we nonetheless did not realize that many species in regional ecosystems stay in stasis together for long periods of time. Likewise, we did not take into account that other major aspect—the larger pattern of coordinated stasis, or turnover pulse—that (as we saw in Chapter 3) is a dominant theme in the history of life over the past half billion years. In other words, *we did not realize that the pattern of species lineage stasis and speciation that we called punctuated equilibria was happening to the vast majority of the species lineages living in close proximity in local and regional ecosystems.*

Stasis and evolution (adaptation through natural selection in conjunction with the speciation process) typically happen simultaneously within many separate lineages of animals and plants, from Cambrian trilobites on up through Pliocene-Pleistocene hominids—the fossil record of our own human evolution. Stability is the norm until a physical environmental event (such as a spurt of global cooling) disrupts regional ecosystems sufficiently that a threshold is reached, and relatively suddenly many species become extinct. Only after such spasms of extinction do we find much evidence of speciation—i.e., actual evolutionary change.

What, then, drives evolution? To my mind, and to many of my colleagues as well, it seems clear that nothing much happens in evolutionary history *unless and until physical environmental effects disturb ecosystems and species.* Competition among genes or organisms for reproductive success is important but in and of itself insufficient to create the dominant patterns we encounter when we examine the history of life.

It is also important to recall Dobzhansky's early message on the hierarchical structure of genetic (i.e., evolutionary) systems. Dobzhansky saw that genes are parts of organisms, which are in turn parts of local populations, which are in turn parts of species. Species, in turn, are parts of larger-scale systems—the higher taxa such as families, orders, and so on, of the Linnaean system. This nesting of genes up through higher taxa, in the functional, dynamic sense of evolutionary theory, constitutes what I call the evolutionary hierarchy—or perhaps better, Dobzhansky's hierarchy.

Ecological systems, too, are hierarchically structured: local ecosystems are connected (through the flow of matter and energy between populations of organisms, each with a different niche) to form regional systems, which themselves are parts of large-scale (e.g., subcontinental) systems, and so forth. All the world's ecosystems are actually joined into one grand global system.

Putting this all together with what we have seen about patterns in the history of life, and with what we have learned about the evolutionary process from the study of genetics and ecological processes yields the outlines of an evolutionary theory that I believe best takes everything into account. Here, in very rough and simple form, is my thumbnail sketch of how evolution works: [18]

On the smallest scales, with little or no environmental disruption and little ecological perturbation, local populations of different species within local ecosystems undergo normal processes of mutation and natural selection, but selection will for the most part be for the status quo. However, different populations of the same species living in adjacent ecosystems will undergo slightly different mutational and selection histories, and in this way genetic diversity within a species as a whole may increase through time.

Ecological disruption of local ecosystems (e.g., damage by fires, storms, oil spills) kills off many individuals within local populations and triggers the normal processes of ecological succession, with pioneer species dominating early assemblages, and species characteristic of later (mature) stages coming in later. Ecosystems are reassembled through recruitment from outlying populations, adaptations already in place are utilized, and little if any evolutionary change occurs.

Longer-term regional ecological disruption (as, for example, when glaciers invade temperate zones from the higher latitudes) disrupt ecosystems even further; in response, species engage in habitat tracking, collapsing toward the Tropics in search of suitable (recognizable) habitat. Yet even in these times of great environmental change, ecosystemic disruption, and displacement of species, natural selection remains dominantly stabilizing as long as species can continue to identify and occupy suitable habitat (which is why many species remain in stasis even during times of momentous environmental stress, as was typical, for example, in the Pleistocene of North America and Eurasia).

Only when environmental stress reaches a threshold—when ecological systems are so severely stressed that they can no longer survive—and when habitat tracking is not an option for many species, does extinction begin to claim many regionally distributed species, clearing the way for rapid speciation events in many separate lineages. This is what is going on at the boundaries between Palmer's Cambrian biomeres, and between the episodes of Brett and Baird's coordinated stasis, and what constitutes, as well, Vrba's turnover pulses (all of which I discussed in Chapter 3). According to the fossil record, most evolutionary change in the history of life occurs in conjunction with these physically induced episodes of ecological stress and extinction. Here, natural selection becomes strongly directional, as new species, with new adaptations, develop rapidly.

Finally, at the grander geographic scales—up to and including the entire Earth—environmental disruption is so severe, and extinction occurs on such a great scale, that entire large-scale arrays of species—taxa such as families, orders, and classes—may go extinct, triggering, as we saw in Chapter 3, the rapid diversification of other lineages, which in many cases are clearly ecological replacements for the lineages that had succumbed to extinction (e.g., see the coral example of Chapter 3).

This is how I—one evolutionary biologist out of many—see the evolutionary process. Microevolution and macroevolution differ only as a matter of scale, as we have seen from the connectedness of all life, and from the sliding scale of events—from the simplest, smallest evolutionary changes up through the enormous effects wrought as the aftermath of global mass extinctions.

I have placed this "sloshing bucket" model of the evolutionary process before you here at the end of this chapter on the evolutionary process not because I insist it is the absolute truth, or anything near the final word on how the evolutionary process works. Quite the contrary: though I suspect most of my colleagues will find much to agree with in this (very abbreviated) sketch of my version of how evolution works, *it is the very strength of the scientific enterprise in general—and of evolutionary biology in particular—that many of my colleagues will not wholeheartedly agree with the foregoing model.*

Creationists claim this is a defect of evolutionary biology—this lack of complete consensus among ourselves. On the contrary: that is its very strength. We have come a long way since Darwin, but we still have a way to go before we can find ourselves in total agreement on all details of how the evolutionary process works. Indeed, realist that I am, I know that day will never come.

We do all agree that life has evolved. We do all agree that the reason why organisms tend to fit their environments so well is that their anatomies, physiologies, and behaviors have been shaped by natural selection, working on local populations living resource-limited lives within the confines of local ecosystems. And we are coming ever closer to agreeing that whatever phenomena of stability and genetic change take place within local populations, gaps between species arise primordially through speciation. It has become, as well, increasingly apparent that speciation events do not take place randomly with respect to one another, but rather are regionally concentrated in bursts that follow close on the heels of environmental disruption of ecosystems and extinction of numbers of species; that is, the data of paleontology are clear on the point that most speciation (and extinction) events involve many unrelated plant and animal lineages, living in association in regional ecosystems, in relatively brief spans of time.

In other words, we agree on a lot of things. But there's still a lot about which we cannot all agree, so work will go on. But it is good, honest work, empirical at base. It is, in short, the very antithesis of the a priori doctrinaire ravings of creationists.

# CHAPTER 5

# Creationists Attack: I
## Scientific Style and Notions of Time

Creationists think that the entire content of the preceding chapters is false—whether maliciously or naïvely or even satanically motivated, in any case thoroughly false. Creationists passionately believe that evolution is the root of much moral decay and downright evil in modern society, and that it must be opposed at all costs. And most creationists still reject the notion that the Earth itself is very old and has had an even longer and equally intricate history as life itself has had.

So what do creationists have to say about the foregoing—the rules of evidence that belong to the empirical-analytical world of science (as opposed, for example, to the rules of evidence and argumentation in a court of law?)[1] Creationists have not yet managed to propose (let alone compile convincing empirical evidence in favor of) a creation model alternative to the well-established patterns and generalizations of geology and evolutionary biology. Indeed, some creationists are willing to admit that theirs is a crusade against evolution prompted by religious beliefs and are often willing to concede that God is by definition supernatural—and therefore not open to empirical investigation in the first place.[2] So the central strategy of all creationist literature has always been to debunk evolution, thereby leaving creationism as the sole standing survivor, constituting the "truth" of the matter. To creationists, it makes abundant sense that, if evolution could somehow be shown to be fatally flawed—to be false—then their version of the history of the cosmos would be instantly established as the correct one.

As far as science is concerned—as the preceding chapters make abundantly clear—though the inevitable disagreements still pervade geology and biology, there are no credible scientific alternatives to an essentially Darwinian view of evolution through natural selection.

So let us pick up the gauntlet and see how well the creationists have mounted their attack on scientific method; on nuclear physics (e.g., in terms of the radiometric dating of rocks); on astrophysics (age of the Universe); on the formation of the geological column and the interpretation of geological history; and, of course, on the evolution of life—the patterns of its history, and what science says it has learned about how the evolutionary process operates. For the moment let's play their game, conceding for the sake of argument that it is indeed an either-or matter, and see who wins on intellectual grounds alone.[3]

By far the closest thing to a clear, concise, and definitive version of what creationists believe to be a rational alternative to the tenets of historical geology and evolutionary biology comes in the formal codifications of "creation science," the form of creationism invented with the explicit goal of injecting creationism into public school curricula under the cloak of science. And though I will use the writings of "creation scientists" as by far the most convenient means of tabulating and analyzing the tenets of creationism as a whole, I will say at the outset that creation science isn't science at all. Creation scientists have not managed to come up with even a single intellectually compelling, scientifically testable statement about the natural world—beyond, that is, hypotheses that have long since been tested and abandoned by science, in many cases as long ago as the nineteenth century. Creation science has precious few ideas of its own—positive ideas that stand on their own, independent of, and opposed to, counteropinions of normal science. And as we shall see, the few ideas that can be construed as scientific are fatuous.

Scientific creationism forms the basis of the bills passed in the 1980s by the legislatures and signed into law by the governors of Arkansas and Louisiana. It is extreme and virulent. It insists that the Earth and all of its life were created in six twenty-four-hour days by the act of a supernatural Creator. Scientific creationists may have failed to contribute anything of substance to the intellectual pursuit—open to all—that is real science, but they have met with some considerable success in promulgating their views in the process of educating U.S. schoolchildren, as well as in the political arena that surrounds that all-important process. And that is reason enough to take a long, hard look at what these people are saying.

## What Are Scientific Creationists Saying?

When debating creationists, scientists are bombarded by a number of challenges that creationists have culled from scientific writings on the natural world. The arguments typically take the following form: "All right, let's see you explain this one!" Hurling challenge after challenge, jumping from atomic physics to zoology, creationists eventually wear the opposition down with their compendia of nature's enigmas. One of their favorites, for example, is the bombardier beetle, darling of creationists because its intricate defense system (bombardier beetles forcefully eject hot fluids when threatened) is impossible (for creationists) to imagine evolving through a series of less perfect, intermediate stages—and, they say, impossible for an evolutionist to prove. Some intrepid evolutionists take them up point by point, and there was a lively exchange on the bombardier beetle in the summer 1981 issue of *Creation/Evolution*, a journal devoted to combating the creationist effort. But such compendia of cases quickly become tedious, and in the end they demonstrate nothing.

Evolutionists admit at the outset that they are puzzled by some of nature's products, and, in any case, as scientists they are in no position to prove anything. But some general themes recur in the literature of scientific creationism—both general objections to what they persist in calling "evolution science," accompanied by countless examples, and statements that make up the corpus of creationist thought on how things have come to be as they are. The bombardier beetle, for example, is but one of the many cases of the general argument that intermediate stages between an anatomical structure and its supposed precursor are impossible, could not be produced by natural selection, and do not, in any event, show up in the fossil record—once again, the hydra-headed problem of gaps.

Other general objections include the argument that apparent design in nature is a prima facie case for a Designer; that complex molecules, anatomical structures, and behaviors cannot have arisen by chance, the probabilities of a natural process forming them being remote; and that the evolution of the complex from the simple violates more fundamental scientific laws—particularly the first and second laws of thermodynamics. To creationists, something cannot come out of nothing (their version of the first law), and once begun, a system inevitably declines and cannot become more complex (their précis of the second law).

These and other sorts of objections, with their myriad examples, are interwoven with various direct pronouncements to form the "creation science" model. Prominent creationist authors such as Duane Gish and Gary Parker (from the Institute for Creation Research, or ICR, in San Diego) have supplied bits and pieces from which we might cobble together a scientific-creation model, but the scientific-creation model given by lawyer Wendell R. Bird in the December 1978 issue of *Acts and Facts* (published by the ICR) is the best place to start, since it provided the basis of the definition of creation science as spelled out in Arkansas Act 590. Bird's list of seven points succinctly summarizes the scientific creationist position, and I will use it as a springboard for discussing all the major creationist claims in the remainder of this chapter and in the following chapters. Bird's scientific-creation model is as follows:

(1) Special creation of the universe and earth (by a Creator), on the basis of scientific evidence. (2) Application of the entropy law to produce deterioration in the earth and life, on the basis of scientific evidence. (3) Special creation of life (by a Creator), on the basis of scientific evidence. (4) Fixity of original plant and animal kinds, on the basis of scientific evidence. (5) Distinct ancestry of man and apes, on the basis of scientific evidence. (6) Explanation of much of the earth's geology by a worldwide deluge, on the basis of scientific evidence. (7) Relatively recent origin of the earth and living kinds (in comparison with several billion years), on the basis of scientific evidence.

Compare Bird's statement with the language of Arkansas Act 590, which defines creation science as follows:

(It) means the scientific evidence for creation and inferences from those scientific evidences. Creation-science includes the scientific evidences and related inferences that indicate: (1) Sudden creation of the universe, energy and life from nothing. (2) The insufficiency of mutation and natural selection in bringing about development of all living kinds from a single organism. (3) Changes only within fixed limits of originally created kinds of plants and animals. (4) Separate ancestry for man and apes. (5) Explanation of the earth's geology by catastrophism, including the occurrence of a worldwide flood. (6) A relatively recent inception of the earth and living kinds.

Quick comparison shows Bird's model and the definition of creation science in the Arkansas law to be virtually identical. The law merely combines Bird's first and third points into one statement (part 1 of the statute) and substitutes a statement about the insufficiency of mutation and natural selection to produce life's diversity for Bird's misuse of the second law of thermodynamics as its second point. All the rest is the same—even the order. Thus Bird's statement of the scientific creationist position is especially important because it has served as the basis for legislation and contains the elements of what creationists would like to see added to the science curricula of all secondary schools in the United States—unless, of course, school boards can be persuaded to drop the subject of evolution entirely from the official curriculum in biology, as happened in the summer of 1999 for the entire state of Kansas. And that is why I choose Bird's statement as a structural guide to my examination—and refutation—of the creationists' nonsensical claims.

### Creationists as Theoretical Physicists

In *Evolution: The Fossils Say No!* (1973), Duane Gish (at the time Associate Director of the ICR and Professor of Natural Science, Christian Heritage College) laments: "The reason that most scientists accept evolution is that they prefer to believe a materialistic, naturalistic explanation for the origin of all living things" (p. 24)—thus anticipating Phillip Johnson's supposedly original attack on philosophical naturalism by some twenty years. But as we have already seen, it is not so much a matter of scientific preference than of necessity: scientists are constrained to frame all their statements in naturalistic terms simply to be able to test them.

When a scientific-creation model such as Bird's, or the definition of such a model enacted as a state law, avers that the origins of the universe, the Earth, and life were the acts of a supernatural Creator, it is automatically excluding itself from the realm of science. (The law changes Bird's "Creator" to "creation," but "creation," especially "from nothing," must directly imply a Creator, as all the creationist literature openly admits.) Gish has admitted, "We do not know how the Creator created, what processes He used, *for He used processes which are not now operating anywhere in the natural universe* [Gish's emphasis]. This is why we refer to creation as special creation. We cannot discover by scientific investigations anything about the creative processes used by the Creator" (1973, p. 40). Thus creationists in effect acknowledge at the outset that creationism in any guise really isn't science at all. One would think that this alone would be enough to keep creationism out of the science curricula of schools. It's simply a matter of definition— of what is science and what is not. By its very definition, scientific creationism cannot be science.

Creation of something from nothing, creationists say, is supernatural. "Matter can be neither created nor destroyed" is the popular cant rendition of the first law of thermodynamics. Creationists like Henry Morris say that evolutionists (meaning, simply, scientists) cheat when they admit they don't know how the universe started. Whence all those particles of matter? Creationists deride the "who knows? maybe it was always there" shrug of an answer that most scientists give. The ultimate origin of matter— beyond any specific theory of a Big Bang origin of the universe and development of the various kinds of atoms, molecules, stars, and galaxies—really is a mystery about which only a few theoretical cosmologists have claimed to have any inklings. Scientists admit when they don't have the faintest idea why or how something happened, and the ultimate origin of matter— at least so far in the annals of science—is a beautiful example. But the creationists know.

The second law of thermodynamics—that all systems tend toward decay and disorder (increase in entropy)—is a great favorite of scientific creationism. Claiming that the Creator first created the Earth and all its life, then set the first and second laws into motion, creationists see the "law of inevitable decline" as the fatal objection to evolution. Evolution is the development of the complex from the simple, they say—precisely the opposite of what you would predict from the second law. Henry Morris (who used to direct the ICR), in *The Scientific Case for Creation* (1977), has a graph (p. 6) showing the creationist expectation: after a perfect beginning, things have gone downhill on Earth—the second law working in its inexorable fashion.

Morris dismisses an apparent exception to the second law: At the moment of fertilization, a human egg is a single, microscopic cell. That cell divides, its daughter cells divide, and so on. The adult human consists of billions of cells. But we are not gigantic multicellular eggs. During development, cells differentiate and take on different forms. They are grouped into specialized tissues, and these tissues are grouped into organs. Creationists admit that adults are, in a basic sense, more complex than the eggs from which they sprang, though, as we have already seen, each cell has the same basic genetic information in its DNA as is contained in the original fertilized egg. To creationists like Morris, however, the process of development from egg to adult does not violate the second law of thermodynamics, because it is only temporary. Death is inevitable, and with it, decay. The second law triumphs in the end.

And, biologists point out, so it will in evolution. It is as certain as anything in science that the sun will not last forever. If, in another 15 billion years, the sun becomes a red giant in its dying days, the Earth will be consumed. In any event, burn out the sun, and the source of our energy is gone. Life will, inevitably, cease to exist. The law will win out. In fact, the vast majority of species that ever lived are already extinct. Some species appear to last for millions of years, but the overwhelming conclusion from the fossil record is that no species can last forever, regardless of what eventually happens to the sun. So, the second law is working here too.

All this is, in a very real sense, merely playing with words, the way the creationists do. Yes, all systems will run down—if there is no fresh input of energy into them. In other words, *the second law applies only to isolated systems*. And the Earth is no isolated system: Plants trap only a fraction of the sun's energy that comes our way; the energy is trapped in the form of sugars, which form the base of the food chain sustaining all animal life. The atmosphere, oceans, and rocky surface of the Earth's crust retain some solar energy in the form of heat. All the rest is reflected back out to space (with a minute amount bounced back again from the moon and artificial satellites). The system isn't running down; it is, instead, open with this continual influx of solar energy. Life uses only a fraction of this available energy; far more is available than is actually used.

In sum, the second law of thermodynamics applies to closed systems—i.e., systems in which no additional energy can enter. In contrast, the developing human embryo—from fertilized egg up through a newborn infant—has a continual supply of energy, from the mother through the placenta. After birth, energy derived from eating and drinking takes over, and in this sense the human body is an open, not a closed, system. The same is true of ecosystems receiving sunlight and sustaining living processes, including the forms of genetic exchange and modification underlying the evolutionary process.

Faced with this argument—that the second law applies to closed, rather than open, systems—Morris and colleagues have simply rewritten the second law, not with the language of mathematics appropriate to the task, but with pseudoscientific jargon. Creationists claim (falsely) that the second law applies to all systems, and they speak mysteriously about systems that require an "energy conversion mechanism" and a "directing program." In the words of Stanley Freske (writing in the spring 1981 issue of *Creation/Evolution*, p. 10), "Creationists are not showing that evolution contradicts the second law of thermodynamics; instead, they are saying that the second law, as accepted by conventional science, is incorrect and insufficient to explain natural phenomena. They insist that something else of their own making must be added—namely a divinely created directing program or a distinction between different kinds of entropy."

Creationists' use of the second law as a general falsification of evolution is a wonderful example of bad science, and (because at first they didn't realize that the law applies only to closed systems) of desperate attempts to salvage their notion. Nothing about this smacks of the scientific; it is, rather, the all-too-human attempt to preserve a pet idea at all costs—even if it requires bending the rules of normal science to serve one's own ends.

## Of Time and the Navel

Creationists say that the universe, the Earth, and all of life are young. All were created within the last few thousand years or so. Henry Morris, R. L. Wysong, and other creationist apologists have devoted many pages to attacking the notion that the universe, the Earth, and life are billions of years old. And they have assembled "scientific evidence" purporting to show that the Earth is only a few thousand years old. It is when they have confronted the rock record that scientific creationists have most vehemently attacked the integrity and judgment of scientists. In their discussions of historical geology, creationists have more than amply demonstrated their capacity for cleverly distorted "scholarship." And it is in their efforts to propound an alternative explanation for the observations of geologists that creationists reveal themselves for the pseudoscientists they really are.

The notion that the Earth has had an extremely long history is one of the great intellectual achievements of human thought. Though there had been occasional flashes of insight throughout history, there was no real need to consider the possibility that the world was vastly older than popularly imagined until the late eighteenth and early nineteenth centuries, when a few men began patiently examining the intricacies of the great sequences of rock strata in Europe. The Greek traveler and historian Herodotus remarked on the seashells he found on Mediterranean hillsides around 400 b.c. Many centuries later, Leonardo da Vinci understood the fossilized shark teeth he found in the surrounding Italian hills to be exactly what they appeared to be: the remains of ancient sharks inhabiting a sea of long ago. But the common conception of fossils up through Renaissance times saw them as petrified thunderbolts—or the work of the devil, who put fossils in the Earth to mislead us all. Today, according to Henry Morris, Satan's role is still perceived in this vein, as it is no less than the devil himself underlying the "well-nigh universal insistence that all this must have come about by evolution" (1963, p. 77).

Creationists accept fossils for what they are: dead remains of once living organisms. Creationists believe that dinosaurs existed, though some think they were wiped out in Noah's Flood, and others contend that Noah had some dinosaur couples on the Ark. In other words, creationists believe that fossils are real all right, but they simply aren't as old as evolutionists insist they are.

When confronted with the evidence that the universe, the Earth, and life really are vastly older than they would like to believe, creationists admit that things certainly *look* old. But, they say, this appearance of old age is illusory. Instead of blaming the devil for tricking us by making the Earth and its living inhabitants seem old (the explanation preferred by their intellectual forebears in the Middle Ages), they blame . . . the Creator! When the Creator made the world, they say, he made it *appear* as if the Earth and life really did have a long history. He had to have done things this way, creationists assert, simply because the universe had to be set in motion: light from distant stars had to have reached the Earth by the end of the sixth day, and rivers had to be already running in their courses. In short, the system had to be up and running at the very moment of creation. But light from distant stars, and streams running in their channels, look as if they've been there a while. It takes time for light to travel interstellar distances or for streams to carve their channels. Often creationists say the Creator's design included a sort of instant history for the universe, the Earth, and life; after Creation, natural processes (such as the laws of thermodynamics) were set in motion, so light continues to reach us from distant heavenly sources, and streams continue to carve their channels.

The idea of evolution had already been "in the air" before Darwin finally published *On the Origin* in 1859. Many people found the notion of evolution disturbing. One such person was Philip Gosse, a clergyman who beat Darwin to the punch with a book of his own, which he called *Omphalos,* the ancient Greek word for "navel," published in 1857. Gosse developed an elaborate argument that it was God's intention to give the Earth a *semblance* of history, just as he gave Adam a navel even though Adam was not born of a woman.

Gosse included fossils in his catalogue of items God created to make us think that the Earth and life are truly old. Many of Gosse's fellow clergymen reacted in horror to this picture of a deceitful God, little better than a nasty devil playing tricks on us. It seemed incomprehensible to most rational minds that a Creator-God would endow humanity with the ability to think, and at the same time take such elaborate steps to fool us. When scientific creationists today claim that the Creator's efforts automatically established an apparent history, they are reverting to Gosse's *omphalos* argument, though of course they deny that the Creator made this mere semblance of history simply to fool us. Rather, they say, it is just the way he had to do it. Either way, it doesn't make a great deal of sense.

What is the evidence that the universe, the Earth, and life are vastly older than the six thousand years Archbishop Ussher (1581–1656) computed from the pages of Genesis? The first inklings came when the Danish physician Niels Stensen (1638–1686, who wrote in Latin and used the Latinized version of his name, Nicolaus Steno) made a few commonsense generalizations about sedimentary rocks in the mid-seventeenth century. Steno saw that most of the layered rocks of the Earth's crust are formed of minute grains of sand, clay, lime, and other mineral substances. He knew that streams carry such particles and discharge them into lakes and seas and that such particles could be observed accumulating in such places in the present day. From these elementary observations, Steno framed the law of superposition: sedimentary beds accumulate from the deposition of particles; the lower beds form first, the upper beds being piled later on top of the lower beds. Thus, beds lying above other beds must be younger than the lower beds.

Creationists do not wholly dispute Steno's law in principle, though their thesis that all the miles of thick sediments were formed during the forty-day Noachian deluge amounts to a rejection of Steno's simple proposition: it would simply be impossible to accumulate the vast thicknesses of sediments deposited since the Cambrian (let alone even earlier geological eras) in a layer-by-layer orderly fashion in such a minute period of time as forty days. Nor do creationists deny that some rocks are metamorphosed (altered by heat and pressure into crystalline form) while others are igneous, cooled from a molten mass such as volcanic lavas. What they *do* reject is the complex chain of observation and reasoning begun in the late eighteenth century and based on the start by Nicolaus Steno.

The late-eighteenth-century physician and farmer James Hutton, as I noted in Chapter 3, essentially founded the modern science of geology when he methodically applied observation, common sense, and a knowledge of present-day processes to explain how the physical features of his Scottish landscape might have formed. Hutton was a Plutonist: he believed that some rocks, such as the lava flow forming King Arthur's Seat, a rock formation at Edinburgh, had cooled from a once molten mass. His opponents were Neptunists, led by the German geologist Abraham Werner (1750–1817), who believed that all rocks, including granites and lavas, schists and sediments, had precipitated out of an ocean that had encircled the globe in primordial times. That controversy died in the first half of the nineteenth century, when it was conclusively shown that some rocks must have cooled from a molten state and that some such igneous rocks (particularly lava flows) lay above older rocks that must have formed from true sediments. Neptunism, as an explanation for the formation of the sequence of rocks in the Earth's crust, is the geological equivalent of the idea of pangenesis, the theory of inheritance adopted by Darwin and other nineteenth-century biologists. Both neptunism and pangenesis are now thoroughly outmoded.

But Hutton prompted another controversy, one that the scientific creationists claim is still alive. Hutton used observations of what was going on in nature around him in the present to interpret the events of the past, just as Steno, Leonardo da Vinci, and Herodotus had done before him. A conflict soon arose: Huttonian geologists saw the action of wind and rain eroding rocks and sending particles rushing downstream to be deposited in lakes and ocean basins as proffering an open vista of long periods of gradual change. Their opponents, such as the French scientist Baron Georges Cuvier and the English clergyman William Buckland (1784–1856), saw it otherwise: called catastrophists, they interpreted geological history as a series of sudden, even violent, happenings, interspersed with periods of quiescence. As I mentioned in Chapter 3, Cuvier thought that the fossil record of life revealed not one but an entire series of separate creations. Buckland saw the physical history of the Earth as a series of cataclysmic events, the last one being the Great Flood of Genesis.

Modern creationists, of course, reject these rather complex notions of catastrophism, calling merely for a single creation of the world and all living "kinds," then catastrophic flood, and then the resultant configuration of things more or less as we see them today. Creationists see themselves as neocatastrophists, and they impugn evolutionists as uniformitarians.

We owe the concept of uniformitarianism primarily to the English geologist Charles Lyell (1797–1875), who followed Hutton's lead and developed a truly coherent science of geology. Lyell, who was to prove so influential on Darwin, yet who rejected evolution until his later years, spoke of the uniformity of geological processes in his famous *Principles of Geology*, published in the years 1830–1833. Creationists today, presumably out of simple ignorance, have utterly misconstrued the modern understanding of Lyell's uniformitarianism. As Stephen Jay Gould and others have pointed out, uniformitarianism meant at least two things to nineteenth-century geologists. It meant that we can seek to understand events of the Earth's past by studying processes of change still going on around us today. It also implied that slow processes, such as the erosion and deposition of sediments, further indicate that all events in Earth history occur at uniform, and usually rather slow, rates. According to this second meaning of uniformitarianism, great changes are the result strictly of the gradual accumulation of minute changes over formidably long periods of time; in this second sense, "uniformitarianism" is a virtual synonym of "gradualism."

The first of these ideas attached to the word "uniformitarianism" is simply common sense—a cardinal assumption if we are to do science at all. The laws of nature that we observe operating today were operative in the past as well. In other words, water has run downhill and the Earth has revolved around the sun ever since there has been an Earth, water, and hills. This is merely the naturalistic assumption, the requirement that all scientific explanations be couched in terms of processes that we have reason to believe are operating in nature, now and in the past—a notion totally counter to the creationist position that different rules were in force at the time of creation, and processes we see operating today were invented by the Creator only after he had created. This first meaning of uniformitarianism is simply another way of stating how all of science is done.

The other meaning of uniformitarianism, however, is another kettle of fish entirely. When creationists say (as they did in Arkansas Act 590) that they are catastrophists while evolution scientists are uniformitarians, they mean that they believe that events of the past were often, if not always, sudden, violent, and cataclysmic. They have the Great Flood specifically in mind, of course. But geologists long ago abandoned this second meaning of uniformitarianism—that all changes in Earth history were the product of infinitesimally minute changes gradually accumulating through time. For well over a century now, we have spoken of the Ice Ages in the recent geological past, when four times huge ice fields have grown over the continents of the Northern Hemisphere. Volcanoes and earthquakes are both infrequent and sudden in their action, and many wreak huge changes. And the invocation of an asteroid impact as the trigger to the ecological collapse that ended the Cretaceous world and wiped out perhaps as many as 50 percent of all living species was a catastrophe par excellence.

Indeed, as we have seen, it is now becoming fashionable to view everything from the evolution of species to the mass extinctions and subsequent proliferations that pepper life's history as a sequence of episodic change—not the slow, steady, gradual change this second meaning of uniformitarianism implies. No, indeed, creationists cannot justifiably claim that they, and they alone, recognize that events in the history of the Earth and its life frequently reflect episodic events rather than slow, steady, progressive change, as Wendell Bird and the Arkansas legislature would have it. Creationists do stand alone as anti-uniformitarians because only they attack uniformitarianism in its first sense: only creationists are willing to suspend natural laws, as we think we understand them today, to frame an ad hoc explanation of the Creator's acts in bringing the universe, the Earth, and life into existence a few short thousands of years ago. But such a stance does not allow them to claim as their own the valid part of the old catastrophism: the notion that the nature of historical events is frequently more episodic than gradual.

## The Geological Timescale

One of the most important weapons in the creationist arsenal is the assertion that the entire scheme of Earth history worked out by geologists is based on faulty logic, or circular reasoning. Henry Morris, for example, says, "Most people do not realize that the very existence of the long geological ages is based on the assumption of evolution" (1977, p. 32), and, "How can the fossil sequence prove evolution if the rocks containing the fossils have been dated by those fossils on the basis of the assumed stage of evolution of those same fossils? This is pure circular reasoning, based on the arbitrary assumption that the Evolution Model is true" (p. 35).

This is the essence of the creationist attack on the notion of a truly old Earth: how do geologists "tell time"? they ask, and then they tell you that, despite what you might think, most rocks cannot be dated directly by measurement of the amount of radioactive decay of various atoms. (This is true.) Instead, they say, geologists use fossils to tell time: they arrange their fossils according to a supposed evolutionary sequence, they correlate rocks all over the world using this supposed sequence, and then they turn around and claim that the fossil record proves evolution. Some creationists have even maintained that when fossils are found out of the "proper" sequence, they are ignored—a charge which is nothing short of a vicious lie. If geologists and paleontologists really were as stupid and self-deceiving as creationists claim, their activities would be as circular and worthless as creationists say they are. The crux of the creationists' argument is that the myth of the geological column and geological timescale is upheld against all contrary evidence by geologists and paleontologists who wish to preserve evolution at all costs. This is a serious charge and is, of course, false.

The truth is that the basics of the geological column—the thick sequence of strata containing the outlines of events in Earth history—as well as most of the basic divisions of geological time, were established well before Darwin published *On the Origin* in 1859. And in fact, to the extent that they held any publicly expressed opinions on the subject, *the geologists who established the basic sequence of divisions of geological time back then were creationists.* The charge that the sequence of subdivisions of geological time—with the Paleozoic, Mesozoic, and Cenozoic Eras as the main divisions of the past half billion years or so—is a ploy to support the false doctrine of evolution is simply untrue. "Paleozoic," "Mesozoic," and "Cenozoic," mean, respectively, "ancient life," "middle life," and "recent life," referring to the animals and plants fossilized in these rocks, the sequence of which was worked out independently of any notion of evolution.

How was our knowledge of the geological column developed? Creationists point out the indisputable fact that in no single place on Earth—even in cases like the Grand Canyon, or the thick sedimentary sequences exposed in such mountain chains as the Alps, Rockies, and Andes—has the full geological column been preserved and exposed. Portions are always missing. In the 1820s and 1830s, explorer and gentleman geologist Roderick Impey Murchison (1792–1871) and Cambridge professor of geology and clergyman Adam Sedgwick set out to study the sequence of strata lying below the Old Red Sandstone, a set of strata widely exposed in the British Isles, now known to be Devonian in age. Using Nicolaus Steno's rule that lower rocks are older than those lying above them, Sedgwick studied the sequence of rocks in Wales from the bottom up. Murchison, meanwhile, working some distance away, was tracing a sequence of layers downward from the Old Red Sandstone. Murchison paid particular attention to the occurrence of fossils in the rocks, while Sedgwick concentrated more on the mineral content.

Both men worked by documenting the physical position of the strata—i.e., which strata overlay which. Each worked out a sequence of position of the rock layers of their areas by examining exposures and noting which beds lay above and below which others. Murchison confirmed that his rocks underlay the Old Red Sandstone; Sedgwick showed that he was studying rocks that lay below Murchison's—rocks that were the oldest of the sedimentary sequence in Wales. Sedgwick called his rocks Cambrian after Cambria, the Roman name for Wales. Murchison called his rocks Silurian for the Silures, an ancient tribe of the region.

The two men, friends and colleagues at first, soon entered into a bitter dispute. As Murchison kept tracing his sequence downward while Sedgwick worked his way up the sequence, they eventually met in the middle; that is, they soon found themselves discussing the same rocks. Each claimed that the intermediate rocks in dispute belonged to his sequence, and the argument was not resolved until 1879, when geologist Charles Lapworth (1842–1920) named all the rocks between Sedgwick's original Cambrian and Murchison's Upper Silurian beds the Ordovician (after the Ordovices, yet another ancient tribe). Today we still recognize the Cambrian, Ordovician, and Silurian as the three oldest subdivisions (periods) of the

THE TRIUMPH OF EVOLUTION

Paleozoic Era. Anyone can still go to southwestern England and Wales and examine the physical sequence of rocks that led to the definition of these three geological periods. It is the physical what-lies-on-what observations that provide the real basis for studying geological time.

Now, it is certainly true that we call some rocks in the United States and elsewhere around the world Cambrian, Ordovician, or Silurian. And it is here that scientific analysis—the testing of hypotheses by seeing if predicted patterns actually occur in nature—comes into play. William Smith (1769–1839), a British surveyor, showed the way. Creationist Henry Morris calls Smith's technique old simply because it was invented almost two hundred years ago (i.e., also prior to the general acceptance of any idea of evolution)—as if the age of an idea has anything to do with its validity. (On those grounds, it is time to get rid of the ideas that the Earth is round and revolves around the sun.)

Smith was surveying the terrain for one of the ambitious canal projects brought on by the Industrial Revolution. Climbing the hills of the English countryside bordering the projected path of the canals, Smith noted that the fossils he saw always occurred in the same order. He could stand on the side of one hill and predict what he would find on the same level on the next hillside, on the basis of his experience with the order of fossils. He could predict, if someone showed him a suite of fossils, which fossils would be found below them and what one could expect to find above them. From this experience, Smith found he could take a mixed collection of fossils and tell the collector, correctly, what the sequence of fossils had been as they lay in the rocks.

There is no assumption of evolution here. It is simple observation: fossils occur in the same general sequence everywhere they are found. When pronouncing two bodies of rock strata—no matter how widely separated they may be—to be roughly equivalent in age (correlative) on the basis of their fossils, there is no evolutionary presupposition whatsoever. The only assumption is that identical, or nearly identical, fossils are the remains of organisms that lived at roughly the same time, wherever they might have lived. This is the basis, for instance, for stating that rocks of Cambrian, Ordovician, and Silurian age occur in the United States as well as in Great Britain, where they were originally studied and named.

We have found Cambrian and Ordovician trilobites and other fossils in the United States—called Cambrian or Ordovician because they are, in some instances, dead ringers for the British fossils. Moreover, these fossils in the United States occur in the same basic sequence as the ones Sedgwick and Murchison found early in the nineteenth century in the rocks of England and Wales. So the correlation of rocks—i.e., saying that two bodies of rock are approximately the same age—is *not* based on the assumption that evolution has occurred. Rather, it is based on simple empirical observation of the order in which fossils occur in strata and on the standard procedures of scientific prediction (i.e., that the same geological sequence will be found around the world) and testing by further observation (which has abundantly corroborated the prediction in over 150 years of subsequent geological field research).

Today we even have an independent means of cross-checking our assumption that similar fossils imply rough equivalence of age for the formation of two or more bodies of rock, for now we have radiometric dating as an independent check.

Geologists quickly saw the potential for the direct chemical dating of rocks soon after Marie Curie (1867–1934) discovered radioactivity. Geochemists, following close on the heels of atomic physicists and chemists, know that unstable nuclei of some kinds of atoms (i.e., isotopes of some elements) emit radiation at statistically constant rates, and in so doing these atoms are transformed into another isotopic form. If we know the original amount of the "parent" and "daughter" isotopes at the time a rock was formed, and if we know the rate at which the parent isotope decays to its daughter isotopic form, we can measure the current ratio of daughter to parent and thus calculate how long ago the rock was in its initial state.

Aha! cry the creationists, lots of assumptions there! How do we know that decay rates are constant? The answer is that laboratory experiments have repeatedly shown that extremes of temperature and pressure fail to alter decay rates,[4] so we continue to have confidence in the theory of radioactive decay—just as we assume that gravity has always been in operation as we observe it today.

Just look at the results. We can take a sample of rock—say, a granite from Nova Scotia. We know it must be Devonian in age because it intrudes rocks containing Devonian fossils but is itself overlain by slightly younger sediments, as judged by the fossil content. Someone from a geochemical lab takes several samples and analyzes the age by three different decay paths between different isotopes of uranium and lead. The ages all come out to be about 380 million years, with a small plus-or-minus error factor of a few million years. (A few million years sounds like a huge error, but a couple of million years one way or the other is a small error compared with the huge age calculated. Saying "380 million years plus or minus 2 million" is like thinking back a year ago and saying you cannot remember whether something happened on the nineteenth or the twentieth of May).

Now, someone from another lab comes along, samples the same Nova Scotia granite, and gets the same results. Then someone else dates a different Devonian granite—one, say, from Greenland—also associated with Devonian fossils. Sure enough, the process works. Rocks predicted to be nearly the same age on the basis of their fossil content always turn out to be nearly the same age when radiometric dates are obtained. And rocks predicted to be older or younger than others always turn out to be older or younger—by the predicted number of millions of years—when dated radiometrically. In short, by now we have literally thousands of separate analyses using a wide variety of radiometric techniques. It is an interlocking, complex system of predictions and verified results—not a few crackpot samples with wildly varying results, as creationists would prefer to have you believe.

Perhaps the most dramatic demonstration of the validity and accuracy of modern geological dating comes from the deep-sea cores stored by the thousands in various oceanographic institutions. The direct sequence is preserved in these drill cores, of course, and the microscopic fossils in them allow the usual this-is-older-than-that sort of relative dating to be done. We can also trace the pattern of changes in the orientation of the Earth's magnetic field: as you go up a core, some portions are positively charged, and others are negative. Major magnetic events, reflecting a flipping of the Earth's magnetic poles, are recognizable, and the sequence of fossils, the same from core to core, always matches up with the magnetic history in the same fashion from core to core.

Then, when we obtain absolute dates from the cores (usually by using oxygen isotopes), we find that the date of the base of the Jaramillo event (one of the pole-switching episodes) *always* yields a date of some 980,000 years ago. The dates are always the same (again, with a minor plus-or-minus factor). They are always in the right order. They are always in the tens or hundreds of thousands of years for the most recent dates, and in the millions of years farther down the cores. There is such a complex system of cross-checking the independent ways of assessing age—all pointing to the same results that I must remind myself that scientists cannot claim to have the ultimate truth.

We have as yet found no rocks directly dated at 4.65 billion years, the estimated age of the Earth. Recall that this is James Hutton's original prediction, as he correctly surmised that the ravages of time preclude the survival of the most ancient crustal materials. The oldest rocks that have been found so far are just about 4 billion years old. The oldest moon rocks, as well as stony meteorites, however, do yield dates of around 4.5 billion years, as already mentioned, agreeing well with the extrapolated age predicted by geochemists of 4.65 billion years for the age of the Earth—a prediction made long before we sent someone to the moon to pick up some samples.

No, the creationists' attack on geological time simply won't work. There are far too many independent lines of evidence—none of which is based on the assumption of, let alone an underlying commitment to, the idea of evolution—that amply confirm what geologists thought must be so 150 years ago: the Earth simply cannot be a mere ten thousand years old. This is no story concocted by a Creator as part of his creative process. The Earth really is extremely old. And, of course, the universe is even older—15 billion years or so, an estimate based on the speed of light and the calculated distance between the center of the universe and its most remote objects.[5] Appearances may be deceiving, of course. The Creator could be only making it look this way. But, leaving a Creator aside as science must, the mundane calculation of modern astronomers is 15 billion years.

But the creationists do not give up. Morris has written that even if the world were as old as geologists say, evolution still would not be proven—which, of course, is correct. Yet most creationists still passionately care that the Earth be proven to be young, and that all the features of the geological record be interpretable as essentially the product of one single event—Noah's Flood. Creationists flatly accuse geologists of covering up the facts to preserve their pet theory of evolution. They point to "polystrate fossils" (their term), by which they mean fossils (usually trees) that are standing vertically and therefore must be sticking up through millions of years of time—if the evolutionists can be believed. Here they pretend that geologists insist that sedimentation rates must always be slow, steady, and even, instead of the truly rapid rates that are sometimes observed. Polystrate trees show every sign of extremely rapid burial, generally when rivers flood over their banks.

But the creationists' favorite ploy to discredit the notion that there is an orderly sequence of rocks and fossils in the Earth's crust lies in their distortion of large-scale rock displacement, which geologists call thrust faulting. Creationists point to areas of the Earth where the fossils seem to be out of sequence, and this is true: in mountain belts, geologists sometimes find older rocks lying on top of younger rocks—in apparent contradiction of Steno's law of superposition. For example, Permian trilobites have been collected way up the slopes of Mount Everest, in rocks lying on top of much younger (Cretaceous) limestones.

Creationists say that a convenient ad hoc explanation—one that they find incredible—is advanced to explain away this "fatal flaw" of historical geology: the concept of massive thrust faulting. Faults are zones where two bodies of rock move past each other. For example, the area of coastline in California west of the famous San Andreas Fault is moving northward relative to the other side of the fault; the San Andreas actually marks the place where the entire Pacific plate is sliding past the North American plate. In other faults, rocks drop down, sliding past other blocks of rock that remain elevated. Africa's famed Rift Valley System is such a place, where the valley floor (with Olduvai Gorge and the Serengeti Plain) is formed by massive blocks of rock dropping down past the higher ground to the west and east. Another kind of fault forms the category in question here: thrust faults occur when regions of rock are crumpled up in the process of mountain building. Sometimes the crumples fracture, and sheets of rock are literally thrust up over other rocks. Large-scale thrust faults are found only in regions of mountain building, where the crust of the Earth has been severely deformed.

Creationists have been uncharacteristically silent so far on the notion of plate tectonics (earlier known as continental drift), a theory that seeks understanding of many features of the Earth in terms of huge slabs of the Earth's crust (plates) changing their position with respect to one another over the course of geological time. For example, peninsular India is reconstructed as part of the Southern Hemisphere supercontinent Gondwana for much of geological time, breaking off only about 70 million years ago and eventually running into Eurasia (about 20 million years ago), and in so doing buckling and thickening the crust and forming the Himalayas. Part of the enormous energy that such processes involve has produced large-scale horizontal movements, in which sections of the Earth's crust have moved many miles laterally. All true mountain belts are folded, like the pleats of an accordion, so mountain belts are all more narrow now than they were as deep basins, when they were accumulating their hugely thick sequences of sediments. It was the crumpling and occasional breakage of those accordion-like pleats of rock that sent Permian strata sliding above Cretaceous rocks on what is now Mount Everest.

Creationists, true to their ways, try to debunk specific examples of thrust faults to show that the whole idea of such faults is an invention by scientists to save their precious idea of an orderly succession of rocks in the geological column. The creationists' favorite example is the Lewis thrust in Montana, where Precambrian (1-billion-year-old) rocks lie on top of fossiliferous Cretaceous rocks only about 90 million years old.

Do paleontologists really invoke overthrusting just to save their story? Are there no independent ways to demonstrate that massive dislocation of strata has occurred? Well, there are—the main one being that, along all faults, in places where both sides of the rocks are exposed, there is (naturally enough) a zone of pulverized rock caused by the scraping of the two rock masses against each other. In addition, slickensides, essentially scars of the movement, are typically seen on the faces of the rock on both sides of the fault. [see Figure 3]

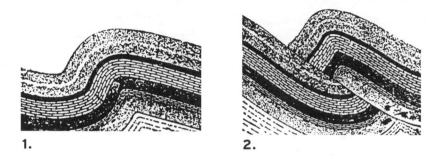

**1.**  **2.**

**FIGURE 3** Development of an overthrust in side view. **1.** Pressure begins to fold a series of rock strata. **2.** With further pressure, the fold breaks, and the rock strata to the right slide over the sequence on the left—leaving older rocks lying on top of younger ones in some parts of the thrust. From Hills, 1953, p. 125, fig. 80.

How do creationists deal with the evidence of thrust faulting? With distortion—and some very poor scholarship. I have a creationist book (J. G. Read's *Fossils, Strata and Evolution,* 1979) that is devoted almost solely to the overthrust problem. In it the author says, in effect, overthrusts are a real phenomenon, mentioning the zones of pulverized rock as tell-tale signs of a real overthrust. So far, so good—though Read fails to mention that the way his "real" examples of overthrusts were first detected was by the anomalous occurrence of fossils, and *not* by the recognition of a thin layer of pulverized rock.

The next step for Read and other creationists (such as Henry Morris) is to turn to what they consider phony examples—the truly massive cases such as the Lewis overthrust, which are the only ones they care about—as they promise to falsify the reality of the geological column in general. Picture after picture (in Read's book) shows the Precambrian rocks sitting over Cretaceous shales in Montana—all without a trace, so they claim, of physical deformation.

What do real geologists have to say about the Lewis overthrust? According to geologist Christopher Weber, who has examined the creationist literature on the Lewis overthrust in some detail, the oft-repeated claim that there is no physical evidence of faulting between the Precambrian and underlying Cretaceous of the Lewis thrust is simply false. Weber (1980) writes:

Whitcomb and Morris [in *The Genesis Flood* (Morris and Whitcomb, 1961, p. 187)] lift the following words from this article [i.e., a professional geological report by C. P. Ross and Richard Rezak, *The Rocks and Fossils of Glacier National Monument*, 1959]: "Most visitors, especially those who stay on the roads, get the impression that the Belt strata (i.e., the Precambrian) are undisturbed and lie almost as flat today as they did when deposited in the sea which vanished so many million years ago."

But, Weber continues, if we read the rest of Ross and Rezak's paragraph, we find that Whitcomb and Morris quoted it out of context:

". . . so many million years ago. Actually, they are folded, and in certain places, they are intensely so. From points on and near the trails in the park, it is possible to observe places where the Belt series, as revealed in outcrops on ridges, cliffs, and canyon walls, are folded and crumpled almost as intricately as the soft younger strata in the mountains South of the park and in the Great Plains adjoining the park to the east."

Even more damning is the thin layer of shale said to occur between the two rock units, evidence of thrusting (as crushed rock) in some areas, but evidence of tranquility (undeformed strata) in the case of the Lewis overthrust, as far as creationists are concerned. Such thrusting, less widespread than creationists would have us believe, and always confined to zones of mountain building where rocks are ordinarily highly disturbed, are not the ad hoc saviors of evolution. Like any other proposition in geology, overthrusts are based on physical evidence, though fossils out of sequence help geologists spot overthrusts in the first place.

## The Noachian Deluge and the Fossil Record

In an exception to their tactic of simply trying to debunk science, creationists have made definitive statements—alternative explanations about how things have come to be as we find them today, statements we can actually test—when using the biblical Great Flood to explain the occurrence of all sedimentary rocks and fossils over the face of the Earth.

Although the only research this notion is said to have directly inspired (as far as I know) was a couple of abortive "arkeological" expeditions to Mt. Ararat, nonetheless the creationist position can be examined on its own merits.[6]

Charles Schuchert (1858–1942), an eminent geologist and paleontologist at Yale University in the early twentieth century, published the *Atlas of Paleogeographic Maps* of North America toward the end of his productive career; the book appeared posthumously in 1955. Thumbing through these maps, anyone can see that today we are in a relatively unusual period of Earth history: the continents today are abnormally dry. The more usual condition by far is for the seas to be flooded over most of the continental interiors. Schuchert's maps reveal a kaleidoscopic pattern of flooding and emergence during the last half billion years as the seas waxed and waned over the continent. Wherever the seas appeared and lasted for some time, they left a covering of bottom sediments; when the seas withdrew for any length of time, erosion would set in and take away the upper parts of the blanket of sediments. And some places—where the seas never reached—never did accumulate a sedimentary record. The upshot: no place on Earth can possibly be expected to have a continuous and complete sedimentary record—of the last ten thousand years, of the last million years, of the last billion years.

So here we have it: a sedimentary rock record, in some places tens of thousands of feet thick (as in the Andes) and in other places totally absent (as over parts of central Canada, where erosion has removed what little amount of sediments ever did cover the granitic core of the North American continent). Geologists explain the uneven distribution of sediments by normal processes of sediment deposition and erosion: deposition where the seas covered the land, erosion when the rocks of the crust are exposed to the atmosphere. Scientific creationists see the entire sequence as the result of one cataclysmic deluge.

Creationists have adopted, with little sign of comprehension, the geological notion of facies. The facies concept points to the simultaneous development of different environments and habitats in different parts of the world. Walking from the seashore inland on the eastern and western coasts of the

United States, for example, takes one from marine habitats, to beaches, then perhaps to lagoons, then to marshes, coastal forests, swamps, and mountains. If all were preserved, each habitat would look different in the rock record, and certainly the kinds of animals and plants preserved as fossils would be different from habitat to habitat. So far, so good. Geologists have been aware of this for years: different kinds of rocks, with utterly different fossil content, may nonetheless be contemporaneous because they were formed in different environments that existed on Earth at the same time.

Creationists claim that this ecological zonation will automatically produce the general order of life that paleontologists have found in the entire fossil record—the fallout of one enormous flooding event. Trilobites, brachiopods, and other invertebrates should appear first: after all, they were already living on the bottom of the sea. Simple, spherical organisms will tend to sink faster than more complex invertebrates, so we have "hydrodynamic selectivity" in addition, to help put the simple creatures at the bottom of the sedimentary pile. Living on land, amphibians, reptiles, birds, and mammals will be buried later by the Great Flood and thus will appear higher up in the rock record, as the seas filled up and encroached on the land. The more clever and advanced the terrestrial animal, the more successful one would predict it to be (according to creationists, that is) in avoiding the calamity of drowning, so dinosaurs are found in lower beds than mammals (actually, not by much!) and humans appear only in the uppermost layers of the sedimentary record.

That is the creationists' answer: ecological zonation, hydrodynamic selectivity, and relative success at fleeing to higher elevations—three points of "explanation" of the sedimentary and fossil record according to the scientific creationist model of a single, worldwide flood. Furthermore, since conditions would have been chaotic during the flood, naturally we would predict some exceptions to the general sequence. This is the cream of creationist pondering over how the fossil record has been formed.

Never mind that the record is, in places, tens of thousands of feet thick, with abundant evidence that much of it (such as limestones and finely laminated shales) must have formed exceedingly slowly. Never mind that careful geological mapping in Colorado and Wyoming, for instance, shows per-fectly clearly how marine rocks—with clams, snails, ammonites, mosasaurs, and other creatures of the Cretaceous briny deep—grade later-ally into terrestrial dinosaur-bearing beds of the same age in Montana. Here is true ecological zonation: Cretaceous animals, both vertebrates and inver-tebrates, living side by side and not piled on top of one another. And here's another example: I am intimately familiar with Devonian rocks in New York State that reveal fish that had occupied streams and ponds and are now pre-served in the present-day Catskill and Pocono Mountains, and that were contemporaneous with shellfish living offshore in the marine waters of western New York and Pennsylvania.

The creationists' argument simply makes no sense. Too many geologists have climbed over those rocks and have seen how they overlie one another. Geologists for the most part don't care very much about the biological sub-ject evolution; that is, they haven't in the past, and as far as I can tell, they still don't care much about it today. What they passionately do care about is the history of the Earth, and of one thing they are certain: the Earth has had a history, a tremendously long and complex history. To disparage the work of geologists over the past two hundred years, to try instead to foist on the naïve the charade that there is no tremendous rock record and that the people who have strived so arduously to understand it are merely fools, is as cavalier an act as I have been sorry to witness. No, it is the creationists—not the geologists—who distort the truth, freely slinging mud at all who cross their peculiarly myopic view of the natural world.

# CHAPTER 6

# Creationists Attack: II
# The Origin and History of Life

The English peppered moth, *Biston betularia*, has been a longtime favorite example in the evolution section of zoology texts, and it has been equally prominent in creationist literature attacking evolution. According to the traditional evolutionary account,[1] this moth species comes in two basic colors: mottled white, and black. The mottled white form beautifully matches the lichens on many English trees; the black moths stand out against the background and are easier targets for moth-hungry birds. During the Industrial Revolution, when pollution from the factories killed the lichens and the trees reverted to the darker color of natural bark, all of a sudden it was the white form that was conspicuous. Black moths soon outnumbered white ones, until comparatively recently, when the crusade against air pollution has once again tipped the scales back in the white variety's favor: the lichens are back in force, and now it is the black moths whose life expectancy is the lower of the two. Here, evolutionists assert, is adaptive evolution—natural selection monitoring environmental change. The moths best suited to prevailing conditions are, on average, more likely to survive and reproduce. The peppered moths provide a beautiful case of small-scale evolution.

It comes as no surprise, then, to find these English moths well represented in creationist literature, too. And it was only a minor departure from their usual course to see that, rather than trying to debunk the example, creationists such as Gary Parker and Duane Gish accept the facts of the moth story—of course, claiming that it somehow supports the creation model.

But I was not prepared to find creationists—particularly Parker and Gish, perhaps the two most eloquent creation "biologists"—actually accepting the moths as examples of small-scale evolution by natural selection! *Modern creationists readily accept small-scale evolutionary change and the origin of new species from old.* That, to my mind, is tantamount to conceding the entire issue, for, as I recounted at some length in Chapter 4, there is utter continuity in evolutionary processes from the smallest scales (microevolution) up through the largest scales (macroevolution).

How can creationists admit that evolution occurs while sticking to their creationist guns and denying that evolution has produced the great diversity of life? Creationists simply insist that the sorts of examples of evolution that biologists give have nothing to do with the wholly new, the truly different. The creationist model is clear on this point: the Creator created "basic kinds," each kind replete with its own complement of genetic variation. Creationists see nothing wrong when they admit that natural selection and reproductive isolation have worked *within* each basic kind, sorting out this primordial variation to produce various specialized types. Creationist R. L. Wysong, a veterinarian, likens the process to the production of the panoply of dog breeds by artificial selection—the great array of different dogs all springing from the same ancestral pool of genetic variation.

Creationists deny that mutations fill the bill as the ultimate source of new variation. Mutations, they claim, are nearly always harmful and are in any case exceedingly rare—precisely the arguments seen as a serious intellectual challenge to Darwinian theory in the earliest days of genetics, until their resolution in the late 1920s and early 1930s. With the advent of molecular biology, genetic variation within species has been shown to exceed by far all previous estimates, and most mutations are small-scale and neither especially harmful nor beneficial when they occur. It fits the evolutionary view of the world that mutations are random with respect to the needs of organisms: mutations don't occur because they help an organism; rather they are mistakes in copying the genetic code—in this sense, no different from the mistakes monks occasionally made when copying medieval manuscripts. That some of these biological mistakes may ultimately prove beneficial is all evolutionists have ever claimed.

## Kinds and Kinds: Creationists and the Hierarchy of Life

If evolution (according to creationists) goes on within but not between kinds, the creationist notion of "kind" becomes rather important. Creationists such as Parker and Gish openly admit that kinds, or basic kinds, are, well, kind of hard to define. The word "kind" has no formal meaning in biology; "kind" to a biologist, if it meant anything at all, would probably mean "species." Here is what Gish has to say about basic kinds; please take the trouble to read his words carefully, for they pose the crux of the creationist position on biological history and present-day diversity and are, at the same time, self-contradictory:

In the above discussion, we have defined a basic kind as including all of those variants which have been derived from a single stock. We have cited some examples of varieties which we believe should be included within a single basic kind. We cannot always be sure, however, what constitutes a separate kind. The division into kinds is easier the more the divergence observed. It is obvious, for exam- ple, that among invertebrates the protozoa, sponges, jellyfish, worms, snails, trilobites, lobsters, and bees are all different kinds. Among the vertebrates, the fishes, amphibians, reptiles, birds, and mammals are obviously different basic kinds. Among the reptiles, the turtles, crocodiles, dinosaurs, pterosaurs (flying reptiles), and ichthyosaurs (aquatic reptiles) would be placed in different kinds. Each one of these major groups of reptiles could be further subdi- vided into the basic kinds within each.

Within the mammalian class, duck-billed platypuses, opossums, bats, hedgehogs, rats, rabbits, dogs, cats, lemurs, monkeys, apes, and men are easily assignable to different basic kinds. Among the apes, the gibbons, orangutans, chimpanzees, and gorillas would each be included in a different basic kind.

When we attempt to make fine divisions within groups of plants and animals where distinguishing features are subtle, there is a possibility of error. Many taxonomic distinctions established by man are uncertain and must remain tentative.

Let us now return to our discussion of evolution. According to the theory of evolution, not only have the minor variations within kinds arisen through natural processes, but the basic kinds themselves have arisen from fundamentally different ancestral forms. *Creationists do not deny the former, that is, the origin of variations within kinds, but they do deny the latter, that is, the evolutionary origin of basically different types of plants and animals from common ancestors* [emphasis mine]. (1973, pp. 34–35)

Gish, of course, cannot possibly mean what he literally says in this passage. He says that "variation" occurs within basic kinds but not between them and proceeds to define such groups as reptiles and mammals as basic kinds. By his very words, then, bats, whales, humans, and the rest of the mammals (and he does acknowledge that human beings are mammals!) he cites could have arisen as variations within the basic mammalian kind. But he then defines these subgroups of mammals as themselves constituting basic kinds, which, according to creationist tenets, means they *cannot* have shared a common ancestor. Bats beget bats, whales beget whales, and so forth, but Gish's words imply that there is no common ancestral connection between these subunits of mammals within the larger, enveloping mammalian basic kind. But if Class Mammalia is "obviously" a basic kind, why can't we see whales and bats as arising simply from variation within a created kind? These statements, of course, are inconsistent at best and nonsensical at worst. One cannot but agree that creationists indeed have trouble with the notion of basic kinds.

What a contrast with the evolutionary position! Evolutionary biologists ever since Darwin have seen that all life is neatly connected within the Linnaean hierarchy: all life, in other words, is intricately nested in ever larger arrays determined by shared genetically based anatomical features. Closely similar species are grouped together, and they form genera with other, only slightly more divergent species; genera sharing unique features form families; families uniquely sharing features form orders, and so forth. All of life fits into this natural arrangement (i.e., not an artificial and arbitrary human construct). And, as we saw at length in Chapter 2, it was Darwin who realized first that a *hierarchical nesting of all living forms must necessarily be the result if evolution—descent with modification—is correct.*

Furthermore, there is no obvious place to draw some sort of dividing line, and to say that smaller groups of species below this line are connected by some sort of natural evolutionary process—but that there are no connections among the larger groups. The reason why Gish himself, try as he might, couldn't do it is simply that it cannot be done. If you admit that dog varieties all belong to one species, and that there is a connection among dogs, coyotes, and wolves, you cannot just stop there; you must also concede (as we have already seen) that, *by the very same token,* dogs share certain similarities with bears and weasels, and these three, as a group, share still other similarities with hyenas and cats, with which they constitute the mammalian Order Carnivora. And, of course, it goes on from there, until all of life falls into this natural organization that is the very imprint of evolutionary history.

Gish points out that "division into kinds is easier the more the divergence observed"—whatever that might truly mean. What is obvious, instead, is that the closer we come to humans, our own species *Homo sapiens,* the smaller the basic kinds Gish and other creationists wish to recognize. The invertebrate groups Gish lists are huge: worms include at least five phyla, snails constitute an entire class of mollusks (comparable at least to the vertebrate classes, such as birds and mammals), and trilobites are an arthropod class. Protoctists (single-celled eukaryotic microorganisms) include many different phyla.

The message is clear: let the paleontologist talk about evolution within the trilobite "kind." Trilobites arose early in the Cambrian Period, some 540 million years ago, and they are last found in rocks approximately 245 million years old. During their roughly 300-million-year sojourn, we know of thousands of species that are classified into numerous families, superfamilies, and orders. But, apparently to creationists, if you've seen one trilobite you've seen them all, and all the changes paleontologists have documented in this important group of fossils are just variation within a basic kind.

I cannot agree. Trilobites are as diverse and prolific as the mammals, and examples of evolutionary change linking up two fundamental subdivisions of the Class Trilobita are as compelling examples of evolution as any of which I am aware. Airily dismissing 350 million years of trilobite evolution as variation within a basic kind is actually admitting that evolution, substantial evolution, has occurred.

But the real reason why creationists care little about trilobites is that they are really worried about only one basic kind: humans. I suspect that creationists would gladly define the rest of life as a single basic kind (and thus allow evolutionary connections among all forms of life) as long as people were singled out as a separate, unique basic kind. After all, Arkansas Act 590 makes a special point of defining a separate ancestry for humans and apes as part of creation science. Yet the degree of biochemical similarity between humans and chimpanzees is greater than 98 percent! It is a source of great satisfaction, I must admit, that with all the attention paid to the biology and fossil record of our own species, as we saw at some length in Chapter 3, it is far easier to demonstrate connections between our own species—the "basic kind" *Homo sapiens*—and the living great apes and fossil hominids, than it is to show connections between the major divisions of trilobites.

We are now in a position to compare the scientific-creation model of the origin of life's diversity with the scientific notion of evolution. Creationists say there can be variation within kinds (microevolution) but not between kinds (macroevolution—"real evolution" to Gary Parker). Biologists assert that there has been one history of life: all life has descended from a single common ancestor; therefore one process—evolution—is responsible for the diversity we see. Creationists insist on two separate theories: (1) the creation of these nebulous basic kinds by a supernatural Creator, followed by (2) microevolution producing variation within those basic kinds. They admit they have no scientific evidence for the first phase.

There is a commonly followed maxim in science (often called Occam's razor) that the simpler idea in general is to be favored over a more complex one when there is no compelling reason to proceed otherwise. The dualistic structure of the creation-science model is a vastly more complicated notion (however barren this structure might be of actual concrete ideas, not to mention evidence) than the simple notion that all life has descended from a single common ancestor—no matter how rich and complex may be the ideas about how that process has worked to produce life's history and present-day diversity.

## Oh, Those Gaps!

Creationists love gaps, lack of any obviously intermediate forms between dogs and cats, insectivores and bats, lizards and birds, fishes and frogs, and so on, and better yet the supposed absence of intermediates in the fossil record. Gaps, to creationists, are the Achilles' heel, the fatal flaw, of biological evolution.

Evolutionary biologists remain unperturbed by the gap problem for several very good reasons: (1) The evidence of connectedness and continuity—whether on a small scale between closely related (often nearly indistinguishable) species, or on larger scales—is simply much better than the creationists claim. In addition, (2) as we saw in Chapter 4, beginning with Dobzhansky's work in the 1930s, evolutionary biologists have come to realize that *the evolutionary process itself—especially via speciation—automatically creates a measure of discontinuity.* And, as Darwin himself noted, (3) many species that would appear as intermediates are now extinct: for example, the australopithecine species, as well as *Homo habilis, Homo ergaster,* and *Homo erectus* are now all extinct, so anyone who would claim close evolutionary connections between chimps and humans (based for example, on their remarkably high percentage of shared genes)[2] nonetheless does not have the benefit of lots of intermediate species, since those intermediates are now extinct. Fortunately, we do have them in the fossil record! Finally, as Darwin also pointed out, (4) we cannot possibly expect to find the remains of all species that have ever lived, for several reasons: the sedimentary record itself is too discontinuous ("gappy"), much has been destroyed by erosion already, fossils tend to be destroyed by chemicals in the groundwaters percolating through rocks (not to mention the ravages of metamorphism), most sediments remain deeply buried and thus inaccessible, and for the most part only animals and plants with tough tissues (e.g., shells, bones, wood) are likely to be fossilized in the first place. With all those things to go against it, the fossil record emerges as a true marvel, and it has produced many series of intermediates, some of which were encountered already in Chapter 3. Let's consider one additional, classic example—one that shows up time and time again in zoology texts and creationist tracts alike: the Mesozoic reptile-bird known as *Archaeopteryx.*

The case of *Archaeopteryx* makes it clear that one person's intermediate is another's basic kind, or failing that, outright fraud. Paleontologists point to examples from their own work, and creationists respond by refusing to accept the examples as intermediates. To evolutionary biologists, *Archaeopteryx* is beautifully intermediate between advanced archosaurian reptiles and birds. In contrast, creationists don't say that *Archaeopteryx* is a fake; to them, it's just another bird. It isn't.

*Archaeopteryx* comes from Upper Jurassic limestones of Bavaria. The seven known specimens are about 150 million years old. Zoologists have known for years that birds are effectively feathered reptiles (dinosaurs, actually), because there are so relatively few anatomical differences between birds and living reptiles, and even fewer between birds and the archosaurian reptiles (including dinosaurs) of Mesozoic times. Birds have some evolutionary specializations not found in reptiles: in addition to their uniquely constructed wings, they lack teeth and have feathers, four-chambered hearts, and horny bills.

Gish says of *Archaeopteryx*, "The so-called intermediate is no real intermediate at all because, as paleontologists acknowledge, *Archaeopteryx* was a true bird—it had wings, it was completely feathered, it flew. . . . It was not a half-way bird, it was a bird" (1973, p. 84). In other words, since evolutionists classify *Archaeopteryx* as a bird, then a bird it is, not some kind of intermediate between reptiles and birds. Semantic games aside, it is certainly accurate to see birds as little more than feathered archosaurs. Feathers, wings, and a bill are three evolutionary novelties that *Archaeopteryx* shares with all later birds, and these new features are the ones that allow us to recognize the evolutionary group birds. But all living birds lack teeth and bony tails, and they have well-developed keeled breastbones to support strong flight muscles. *Archaeopteryx* lacks such a keel but still retains the teeth and bony tail typical of its reptilian ancestors.

The reason why *Archaeopteryx* so delights paleontologists is that *evolutionary theory expects that new characteristics—the "evolutionary novelties" that define a group—will not appear all at the same time in the evolutionary history of the lineage.* Some new characters will appear before others. Indeed, the entire concept of an intermediate hinges on this expectation. Creationists imply that any intermediate worthy of the name must exhibit an even gradation between primitive and advanced conditions of each and every anatomical feature. But there is no logical reason to demand of evolution that it smoothly modify all parts simultaneously. It is far more reasonable to expect that at each stage some features will be relatively more advanced than others; intermediates worthy of the name would have a mosaic of primitive retentions of the ancestral condition, some in-between characters, and the fully evolved, advanced condition in yet other anatomical features. *Archaeopteryx* had feathered wings, but the keeled sternum necessary for truly vigorous flight had not yet been developed in the avian lineage. And *Archaeopteryx* still had the reptilian tail, teeth, and claws on its wings.

Creationists point to some living birds that, while still young, have poorly developed keels or claws on their wings (as is the case with the South American hoatzin). They also point to Cretaceous birds, younger than *Archaeopteryx*, that still had teeth. Here is the height of twisted logic: creationists say, "Look here—there are some modern and fossil species of birds with some of the supposed intermediate or primitive reptilian features that are out of the correct position in time." Instead of interpreting these birds as primitive links to the past, creationists see them as somehow a challenge to *Archaeopteryx* as a gap-filling intermediate.

The whole point about intermediates, though, is that ancestral features are frequently retained while newer features are being added to another part of the body. It was not for another 80 million years or so that birds finally lost their teeth—though they had lost their tails in the meantime. That juvenile stages of descendants often show features of their adult ancestors, as in the hoatzin's juvenile claw, prompted the German evolutionary biologist Ernst Haeckel's (1834–1919) famous nineteenth-century maxim "ontogeny recapitulates phylogeny," meaning that the evolutionary history of an animal is in a sense repeated in its development from egg to adult. Bluster as they might, creationists cannot wriggle away from *Archaeopteryx*.

## Creationists Confront Human Evolution

The case the creationists care about most—and quite possibly the *only* case they care about—is the origin of humans, *Homo sapiens*, us. And it is indeed

ironic that delineating humans as a basic created kind separated by profound and unbridgeable gaps from the living great apes and extinct species
of hominids is a task of Herculean proportions, a challenge that so far has
evoked only a feeble response from creationism's leading exponents.
Creationists have had pitifully little to say about this, their worst nightmare:
the overwhelming genetic and anatomical evidence of connections between
humans and the great apes (and, through the apes, with the rest of life), and
the dense and rich fossil record of human evolution.

What do the creationists say about human evolution? Creationists such as
Gish and Parker agree with anthropologists that younger fossils are very
modern in appearance, though they don't admit that anatomically modern
human fossils from southern Africa are over a hundred thousand years old,
and that modern humans arrived in Europe around thirty-five thousand
years ago. They also like to revive the old canard that, with a necktie and coat
on, a Neanderthal man would pass unnoticed in the New York subways. I
sincerely doubt it, since most paleoanthropologists have generally concluded that Neanderthals were a distinct species.

Skipping back 3 million years or so, we find various species of the genus
*Australopithecus,* whose name means (as creationists fondly point out)
"southern ape" and thereby, on linguistic grounds alone—automatically, in
the creationist book—is a form of ape and no member of the human lineage. Assessing zoological relationships on such etymological grounds is
rather dubious, to say the least, but the creationists' claim that "it looks like
an ape, so call it an ape" greatly insults these remote ancestors and collateral kin of ours. They had upright posture and a bipedal gait, and some of
them, at least, fashioned tools in a distinctive style. No apes these, but rather
primitive hominids looking and acting just about the way you would expect
them to so soon after our lineage split off from the line that became the modern great apes.

But it is the fossils of the middle 1.5 million years I just skipped over that
make creationists writhe. Here we have *Homo erectus,* first known to the
world as *Pithecanthropus erectus* (literally "erect ape-man"), based on specimens from Java, and as *Sinanthropus pekinensis* ("Peking man") from China.
Now known from Africa as well (in the form of the closely related species,

*Homo ergaster*), the *Homo erectus* lineage lived virtually unchanged for over 1.5 million years and was, by all appearances, a singularly successful species. *Homo erectus* had fire and made elaborate stone tools, and its brain size was intermediate between that of the older African fossils and the later, modern-looking specimens. Specimens of *Homo erectus* don't look like apes, yet they don't look exactly like us, either. To most of us, *Homo erectus* looks exactly like an intermediate between modern humans and our more remote ancestors.

What do creationists do with *Homo erectus?* No problem: *Homo erectus* is a fake in the creationist lexicon. Gish asks us to recall Piltdown, that famous forgery—evidence of skulduggery in the ranks of learned academe. And what, the creationists ask, about *Hesperopithecus haroldcookii* ("Nebraska man"), described years ago on the basis of a single tooth, which later turned out to have belonged to a pig? (Scientists do make mistakes, and pig and human molars are rather similar, presumably a reflection of similar diets.) So it is, they say, with *Homo erectus.* According to Gish, the Java fossils were just skullcaps of apes wrongly associated with a modern human thighbone. And the original Peking fossils are now gone, apparently lost by a contingent of U.S. marines evacuating China in the face of the Japanese invasion of World War II. Hmmm, very suspicious, say the creationists. Never mind the casts of the originals, the drawings and photographs plus detailed written descriptions of these fossils published by the scholar Franz Weidenreich. And never mind that the Chinese have since found more skulls at the original site, or that Richard Leakey has found the closely related species, *Homo ergaster,* beautifully preserved in East Africa.

That the best the creationists can do with the human fossil record is call the most recent fossils fully human, the earliest merely apes, and those in the middle—the intermediates, if you will—outright fakes is pathetic. Humans are about the worst example of a basic kind that creationists could have chosen. The irony is great: the case toward which all their passion for producing propaganda is ultimately directed—how *we* got here—is about the most difficult one I can think of to support the model of creation.

## Creationists on the Fossil Record: A Final Note

It would certainly be helpful to the creationist cause if all organisms could be shown to have appeared at the same time in the rock record—the result of one grand creative act by God—or, failing that, at least a grand commingling of extinct and modern forms of life all deposited together by Noah's Flood. Thus creationists have taken great delight in the supposedly human footprints alongside bona fide dinosaur tracks in the Cretaceous Glen Rose Formation, exposed in the channel of the Paluxy River in Texas. Here, they proclaim, is direct evidence that humans and dinosaurs roamed the Earth together, just as it was written in *Alley Oop*. Gary Parker is quite suave as he describes fitting his own feet into these impressions. But none other than a creationist (B. Neufeld, in his article "Dinosaur Tracks and Giant Men," 1975) has blown the whistle on these tracks. Alas for the creationist cause, they aren't footprints at all; the few "human" impressions visible these days do not show any signs of "squishing" of the sedimentary layers either at the edges or directly beneath the "tracks" (as the dinosaur prints, incidentally, clearly do). And, according to Neufeld, during the Great Depression it was a common local practice to chisel out human footprints to enhance tourist interest—a practice akin to the recent fabrication of Bigfoot footprints in the Pacific Northwest. Need anything more be said about the quality and trustworthiness of creationists' dealings with the fossil record?

## Sparring with Luther Sunderland

Earlier I said that creationists are poor scholars at best and at worst have been known to distort the words and works of scientists. Throughout the creationist literature, one sees repeatedly statements such as this from Gary Parker:

> Famous paleontologists at Harvard, the American Museum, and even the British Museum say we have not a single example of evolutionary transition at all. (1980, p. 95)

This statement is untrue. We have already encountered Luther Sunderland as the creationist who lured me into confrontation in the first place (see Chapter 1) and the man who bragged that he convinced Ronald Reagan's speechwriters to inject a bit of skepticism about evolution into a presidential election campaign. Sunderland interviewed prominent paleontologists at various museums and universities. I was one of them. Some of us tried to discuss some procedural difficulties in recognizing ancestors,[3] also admitting that the fossil record is full of gaps. Nothing new there. All the paleontologists interviewed later told me that they certainly did cite examples of intermediates to Sunderland.

Sunderland then wrote letters to newspapers and testified in various venues (e.g., to the Iowa State legislature, as already mentioned) that the paleontologists he interviewed admitted that there are no intermediates in the fossil record. In 1984, Sunderland published *Darwin's Enigmas: Fossils and Other Problems*. Later he went on to write *Darwin's Enigma: Ebbing the Tide of Naturalism*, copyrighted in 1988 but first printed in August of 1998. Sunderland says that his 1998 book "presents the substance of these interviews through the use of short excerpts and summaries of the replies to the questions" (p. 13).[4] It is worthwhile taking a look at what Sunderland is up to here, and I'll do so by picking out the most egregious thing he has me saying—or has to say about me. My actual opinions on all the other distorted issues with which Sunderland saddles me (and my paleontological colleagues) in the course of his book are, of course, revealed in the pages of this book, though not in the format of a rejoinder specifically to the work of Luther Sunderland. I will single out here the worst case—one that creationists are still using in their writings and debates (including on the Internet), and one that has defenders of evolution wondering if I *really* said what Sunderland has me saying in his book.[5]

The issue, once again, is gaps—the supposed lack of intermediates in the fossil record—and revolves around what I purportedly said about horse evolution or, rather, about an exhibit on horse evolution that had been on continual display at the American Museum of Natural History for many years. Photos of this exhibit were routinely reproduced in zoology textbooks for much of the twentieth century. Here is Sunderland's version of what I said to him about that exhibit:

> I admit that an awful lot of that has gotten into the textbooks as though it were true. For instance, the most famous example still

on exhibit downstairs (in the American Museum) is the exhibit on horse evolution prepared perhaps fifty years ago. That has been presented as literal truth in textbook after textbook. Now I think that that is lamentable, particularly because the people who propose these kinds of stories themselves may be aware of the speculative nature of some of the stuff. But by the time it filters down to the textbooks, we've got science as truth and we've got a problem. (1998, pp. 90–91)

So there you have it: Sunderland has me slamming my curatorial predecessors at the museum for misleading the public. A few pages later, Sunderland gets back to me and the horses and makes a serious charge about *my* integrity; please bear with this rather lengthy quotation, as it reveals the heart and soul of creationist tactics:

When scientists speak in their offices or behind closed doors, they frequently make candid statements that sharply conflict with statements they make for public consumption before the media. For example, after Dr. Eldredge made the statement about the horse series being the best example of a lamentable imaginary story being presented as though it were literal truth, *he contradicted himself.* The morning of the beginning of the Seagrave's [sic] trial in California he was on a network television program. The host asked him to comment on the creationist claim that there were no examples of transitional forms to be found in the fossil record. Dr. Eldredge turned to the horse series on display at the American Museum and stated that it was the best available example of a transitional sequence. On February 14, 1981, Sylvia Chase, host of the ABC television program "20/20," questioned him on this subject as follows:

Sylvia Chase: "Dr. Niles Eldredge, Curator of the Department of Invertebrates of the American Museum of Natural History, is one of many scientists vigorously opposed to the creationists. I asked him for evidence (of evolution)."

Dr. Eldredge: "Ahh, the horse is a good example. Here is an effectively modern horse which is a million years old, but we can all recognize it as a horse. And as we go deeper in lower layers of rock, back further in time, we excavate successfully more primitive horses. Here's one that is two-million years old. They are becoming progressively less and less obviously horse-like till we get back 60 million years ago, and here is the ancestor of the rhinoceros—or very close to the ancestor of the rhinoceros.[6] So that when the creationists tell us that we have no intermediates between major groups, we point to a creature like the dawn horse and say 'Here we have an exact intermediate between horses and the rhinos.'" *So, in 1981, after joining the anti-creationist campaign, Dr. Eldredge repeated a scenario for a nationwide audience that in 1979 he had called "lamentable"* [italics mine]. (pp. 94–95)

Now, I have no idea whether the words Sunderland puts directly in my mouth are accurate, verbatim accounts of what I actually said; some of this doesn't sound particularly like me, and after all, I have only Luther Sunderland's word for it—and he is, in effect, calling me a liar. But let's assume I actually said everything that he quotes me as saying in these two passages. If so, the reader might well be inclined to agree with Sunderland—that I was talking out of both sides of my mouth, blatantly contradicting myself—and even believe that I did so for the ideological purpose of defeating creationism, after I had joined the anticreationist campaign.

Well, there is no doubt that I am a thoroughgoing anticreationist, and one of the major reasons is the falsely malevolent light in which creationists cast evolutionary biologists: if we are not just plain stupid fools, then we are liars who say one thing to each other and quite another to the world at large.[7] But here, for the record, was what I was talking about—first with Luther Sunderland, and then with Sylvia Chase:

Eldredge with Sunderland (1979): As codeveloper with Stephen Jay Gould of the theory of punctuated equilibria, I was very sensitive to any and all claims made, past and present, by my paleontological brethren to the effect that evolution is a phenomenon of slow, steady, gradual, and progressive change *both within species through time and also between successions of species through time.* My problem with the old horse exhibit is that it depicted horse evolution as linear and gradual, though I must also say that nothing in the old text said anything whatsoever about gradual evolution. Still, it was a fair inference that that is what the take-home message would be. Remember, this was still the early days of trying to get paleontologists to reexamine their data, and to see whether or not species—of horses, dinosaurs, corals, whatever—were as typically stable as my trilobites. I am, in this connection, extremely gratified to report that, unsurprisingly, now that the work has been done (post-1979), the predominant patterns of horse evolution fit the patterns of stasis followed by extinction and speciation that I have discussed (in Chapters 3 and 4) as being utterly typical and general for all of life. In short, my outburst to Sunderland was on the subject of evolutionary gradualism not on the question of anatomical intermediates.

Eldredge with Chase (1981): The dead horse that Sunderland and all other creationists beat is, of course, not stasis versus gradualism, but the existence of anatomical intermediates, especially if they exist in perfect stratigraphic order. I am here to tell you that my predecessors had indeed unearthed and mounted a wonderful series of skeletons, beginning with the Eocene *Hyracotherium* (the so-called dawn horse), with its small size, four toes on the front feet, five on the back feet, shortened face, and generalized perissodactyl teeth suitable for browsing, not grazing. Climbing up the Tertiary stratigraphic column of the American West, we find the horses becoming progressively bigger, with fewer toes (modern horses have but one on each foot) and more complicated teeth. The horses of the Pliocene are essentially modern.

*This is not a made-up story. Those fossils are real. They are in the proper order, and they are a spectacular example of anatomical intermediates found in the exact predicted sequence in the rock record. They are every creationist's nightmare.*

No, horse evolution was not in the straight-line, gradualistic mode. But to state or imply that the horse evolution exhibit was somehow arranged to support an evolutionary story—to imply that the old museum curators deliberately misled the public by arranging the order of these horse fossils as they saw fit—is a damn lie.

The upshot here is that the fossil record of horses is now known to be many times more dense and richly diverse than in the days when that old exhibit was first mounted. Yes, there are many side branches, and stasis, rapid evolution in speciation, and turnover pulse–related phenomena (such as extinction) are as utterly typical of horse evolution as they are of all other forms of life that have left fossil records behind. If anything, we know more intermediate anatomical forms in horse evolution than we did when that fabulous old exhibit was first mounted.

Creationists hear what they want to hear because they believe what they want to believe. They obviously think that all is fair in both love and war, and they see this as a culture war. But somehow I persist in the apparently quaint belief that lying, cheating, and distortion are inherently unchristian.

### In the Presence of a Lawyer: The Case of Phillip Johnson [8]

Phillip Johnson is, hands down, the most visible and successful creationist of the 1990s and early part of the new Millennium. Boalt Professor of Law at the University of California, Berkeley, early in his career Johnson clerked for Chief Justice of the United States Supreme Court Earl Warren. He is, in short, no dummy, and in the one "debate" I had with him (along with a TV interview and classroom and faculty discussion sessions—not to mention the obligatory postperformance drink)—at Calvin College in Grand Rapids, Michigan, in January 1996—I found him generally affable.

Johnson lectures widely; he also writes rather well, and I gather that his several books and many articles enjoy a wide circulation and readership. A darling of the Christian right, Johnson believes strongly that what he sees as a culture war between atheistic philosophical naturalism and what he calls theistic realism is manifested in the loss of esteem and respect for theology on university campuses. Despite directing a substantial proportion of his efforts at Christian laity, perhaps above all else Johnson seeks intellectual respectability within academe.[9] This is indeed something new in creationism, and Johnson's positions need to be examined: What, exactly, is he saying? And how much of his thinking is really new?

Johnson makes no bones about either his born-again Christian beliefs or his conservative political views, and this alone, I must say, I find refreshing after decades of combating wolf-in-sheep's-clothing scientific creationists who insisted (still do, for that matter) that theirs is actually a scientific position—rather than a religiously inspired one—and, as such, deserving a place in the science curricula of public schools. The first of Johnson's books, *Darwin on Trial* (1991), however, reveals relatively little of either Johnson's religious and philosophical background, or of what have later emerged as his main arguments. In fact, *Darwin on Trial* is really little more than a straightforward (though up-to-date) standard creationist antievolution tract—one that is curiously, yet one suspects purposively, divorced from other works of that genre.[10]

In most of his earlier public presentations, Johnson spent most of his time attacking the fossil record—the old question of gaps and intermediacy. He made many appearances with Cornell historian of science William Provine, for example, and a publicly available tape of one of their joint presentations (at Stanford University in 1994) shows Johnson performing a not overly clever attack on the lack of intermediates, as especially revealed in a museum display devoted to the Cambrian explosion. I mention this instance simply to reiterate Johnson's fundamental stripe as a fairly run-of-the-mill antievolutionist for the most part. What is different about the presentation from one given, say, by Duane Gish, is that Johnson, in the rebuttal period, finally gets down to some of the details of what really is somewhat novel about his approach: his insistence that science in general—and evolutionary biology perhaps in particular—reflects an underlying philosophical stance he calls philosophical naturalism, which inherently, by its very definition, is atheistic. On the tape, admitted atheist Provine wholeheartedly agrees—about which, more anon.

Phillip Johnson believes in a personal God—a God who is all-powerful and has, in fact, created everything we see around us. His is a decidedly proactive God, taking part in causation of things small as well as grand. Johnson has said repeatedly that, if evolution is true, then God is thereby reduced to a do-nothing, boring kind of God. Put another way, science at the very least marginalizes God: "The acceptance of naturalistic assumptions in science by Christian and secular intellectuals alike has moved God steadily into some remote never-never land ('before the big bang') or even out of reality altogether" (Johnson, 1995, p. 111). Actually, it is the "out of reality altogether" that Johnson really has in mind when he says that science *assumes* that all that exists is the material world and the forces that hold it together and cause its various elements to interact.

But, Johnson says, *if there really is a God that is shaping the Earth and creating all life, then science ought to want to know about that.* After all, as Johnson says, if science is about understanding the natural world, and if God has a direct hand in shaping that natural world, then why in the world would science choose to ignore—nay, even deny—God's very existence? In assuming that the natural, material world is all there is, scientists and their meek followers are automatically ruling out what is perhaps *the* crucial element in shaping the world! And that, in brief, is Phillip Johnson's philosophical naturalism: the assumption that the material, physical, natural world—with its bits and pieces and its characteristic modes of interaction—is really all there is out there.

Johnson's second book, *Reason in the Balance* (1995), says all this at great length, but it really doesn't try to come to grips in any detail with what "naturalism" really means until the appendix. Johnson wrote a paper prompted by "a remark by a Christian college professor who had argued that my 'creationist bias' was affecting my assessment of the scientific evidence for evolution. I include the paper here as an appendix instead of trying to fit it into the text, because the issues that fascinate persons who devote a professional interest to this subject may be overly complex for general readers who have other matters to occupy their attention. On the other hand, I want to preserve this statement as a starting point for further discussion among professional academics in particular" (p. 206).

Well, I won't try to hide my answer to Johnson's tirade against philosophical naturalism in an appendix. Here, in a nutshell, is what is wrong with Johnson's argument: his dichotomy between philosophical naturalism and theological realism. It is the answer I conceived when I accepted the offer to debate Johnson, and it is the same answer everyone else has reached.[11] Unlike Johnson, I do not see these issues as overly difficult for anyone to grasp.

Everyone—even Phillip Johnson—agrees that there is a physical, material world. Everyone also agrees that there is something called human knowledge, and that human knowledge has grown over historical time. *Science is a way of knowing about the nature—composition and behavior—of the natural, material world.* That's not nothing, but that is all science is: a set of rules and an accumulated set of ideas, some more powerfully established than others, about the nature of the material world. *By its own rules, science cannot say anything about the supernatural.* Scientists are allowed to formulate solely ideas that pertain to the material universe, and they are constrained to formulate those ideas in ways that can be testable with empirical evidence detectable by our senses.

Johnson says that restricting analysis purely to material, naturalistic terms is automatically atheistic—amounting to a de facto claim that God does not exist. But science does not—because it cannot—say that only the natural material world exists. Rather, science is restricted by the limitations of human senses and was, in any case, invented solely to explore the nature of the material universe. It does not rule out the existence of the supernatural; it merely claims that it cannot, by its very rules of evidence, study the supernatural—if, indeed, the supernatural exists.

Johnson, naturally enough, loves scientists who agree with him—scientists (and historians of science like Provino) who are all too eager to announce that science and religion are truly at loggerheads, that science implies there is no God. Once again, he uses the either-or approach: either you believe that God exists and fashioned the world we find, or you believe that the material world is all that exists and that by definition there is no God.

Johnson—like Gish and so many other creationists—intensely dislikes what the old-line creationists simply called theistic evolution, what Johnson prefers to call theistic naturalism. Theistic evolution (or theistic naturalism) is the position that God created heaven and Earth, and all manner of beast, including humans, but did so using natural laws. Johnson and other hard-line creationists find this line of argument unacceptable because it relegates God to a sort of caretaker status, more or less content to sit on the sidelines. So Johnson simply dismisses theistic naturalism, even though it has been the line of reasoning of choice of mainline Protestant denominations since the nineteenth century—one that many, many devout people (including some scientists as well as nonscientists) still profess.

Creationists are famous for their basic strategy of debunking evolution rather than proposing their own model, testable or not, on how God fashioned the diversity of life and breathed life into humanity. Johnson has been steadfast in this approach for most of his career, but he does, in *Reason in the Balance,* offer a brief characterization of his theistic realism. Johnson begins by stating, "If theologians hope to win a place in reality, however, they have to stop seeking the approval of naturalists and advance their own theory of knowledge. My intention here is to start the process, rather than finish it, but readers are entitled to expect me to provide a concrete proposal as a basis for further discussion" (1995, p. 107). After citing the Gospel of John to reveal "the essential, bedrock position of Christian theism about Creation," Johnson then concludes this brief section on theistic realism with the following statement: "If Christian theists can summon the courage to argue that preexisting intelligence really was an essential element in biological creation and to insist that the evidence be evaluated by standards that do not assume the point in dispute [Johnson means here philosophical naturalism], then they will make a great contribution to the search for truth, *whatever the outcome*" [Johnson's emphasis] (p. 110).

Well, we might ask, is there any *evidence* of God's direct participation in the formation of the Earth and the creation of all life? Johnson thinks so, and he is happy to parade the work of some of the members of his inner circle. It turns out, though, that theirs is strictly the up-to-date version of the same old creationist arguments for seeing the history of life as solely the outcome of God's direct handiwork.

### Design, Chance, and Complexity

Creationist authors have devoted entire books to their interpretations of the data of biology. But, as has already become abundantly clear, apart from their convoluted arguments about fossils and basic kinds, they find relatively little in biology that they can offer as supportive of the creation model—known to Phillip Johnson as theistic realism. What little there is of substance (if it can be called such) centers on the notions of design, chance, and complexity.

As briefly mentioned in Chapter 1, one of Darwin's first and most persistent critics after *On the Origin* appeared was St. George Mivart.[12] Mivart hounded Darwin on a problem with which he was already amply troubled: how could one imagine a structure as complex and beautifully suited to perform its function as a human eye to have evolved through a series of simpler, less useful and efficient stages?

Anatomists were among the last holdouts against accepting the idea of evolution, so entranced were they with the intricate complexities of the organ systems they studied. Imagining intermediate stages between, say, the front leg of a running reptile and the perfected wing of a bird seemed to them impossible, as it still does to today's creationists. That the problem perhaps reflects more the poverty of human imagination than any real constraint on nature is an answer not congenial to the creationist line of thought. In the taped debate between Will Provine and Phillip Johnson already cited, for example, Provine lists intermediates between climbers and fliers, alleging (correctly), for example, that flying squirrels can soar like mad without relinquishing their abilities to scamper and cling to trees. Johnson, of course, never bothered to respond.

Thus the complexity argument is just a subset of the creationist claim that there are no intermediates. So naturally, and by the same token, creationists reject any evolutionary biologist's claim that, for example, there are indeed intermediate stages between simple eyes with a few cells covered by a simple lens up through more complexly configured eyes (it is noteworthy that they always say the human eye, when the human eye is configured in essentially the same way as any other mammalian eye).

The latest manifestation of the creationist argument on complexity is in the writings of one Michael Behe, a biochemist at Lehigh University. Once again, it is old wine in new bottles. In the one "debate" I had with Behe, he grudgingly acknowledged that such time-honored conundrums as the evolution of the vertebrate eye have been, in fact, effectively resolved by evolutionists. And in his book *Darwin's Black Box* (1996), Behe says that Darwin "succeeded brilliantly"—not by "try[ing] to discover a real pathway that evolution might have used to make the eye. Rather, he pointed to modern animals with different kinds of eyes (ranging from the simple to the complex) and suggested that the evolution of the eye might have involved similar organs as intermediates" (p. 16), and he goes so far as to supply a diagram of three different eyes of varying complexity.

But the real problem, according to Behe, is not so much the anatomical structure of the human eye as the problem of vision itself. The "irreducible complexity" on which he prefers to concentrate lies in the molecular (chemical) level and in general "refers to a single system composed of several well-matched, interacting parts that contribute to the basic function, wherein the removal of any one of the parts causes the system to effectively cease functioning" (p. 39). Anatomically speaking, an eye might struggle along with removal of one or more of its parts, but at the molecular level Behe swears that such is not the case, and there is no way that mutation or selection could have assembled the intricate, complex molecular pathways that underlie the physiological process of vision.[13]

Note that what Behe has done is simply push the problem back one more notch: same problem, just at a different level. There is really little reason to believe that the evolutionary pathways leading to particular molecular reactions—underlying vision or anything else—will never be completely understood. And *that* takes us to the ultimate degree, for the more intermediates paleontologists and anatomists find, the more recalcitrant creationists become: they just *won't* believe, because they already believe something else. There really is *nothing* different about Behe's argument from any other use of the argument of complexity over the past 150 years.

Interwoven with the difficulty in imagining the gradual evolution of complex organs are two separate themes: (1) the more complex a structure is, the more eloquent a silent argument it is for the conscious work of a Designer, and (2) the more complex a structure is, the more improbable it is that it arose by chance alone.

The argument that nature is so complexly organized, with each creature specially suited to the role it plays in the economy of nature, that only a Creator could have fashioned things in this way is an old one. It was the particular view of the theologian-naturalists prior to *On the Origin,* and it is still in use today in the creationist literature. Creationists usually talk of watches, though somewhat refreshingly, creationist Gary Parker (in his book *Creation: The Facts of Life*, 1980) prefers to use Boeing 747 jumbo jets as his example. The "argument" is simply that such complex machines, so admirably suited to the purposes they serve, require a watchmaker or an elaborate assembly line of airplane builders, respectively. All the parts must be premeditatedly put together by expert craftsmen. Alone, no spring or jewel can keep the time (this argument is no different from Behe's "irreducible complexity"). Only when the watchmaker cleverly arranges the parts in precisely the right way does the watch become functional. Clearly, the very existence of watches directly implies the existence of a watchmaker. So, too, creationists argue, does the existence of complex organisms imply a conscious Creator.

Now, as a scientist, I'll grant that a Boeing 747 implies a creator. I've seen pictures of the assembly line, and more to the point, I am aware that the aluminum of which the airplane is made is extracted (with great difficulty and expenditure of energy) from its complex ore—a process known only, insofar as I am aware, to human beings. I will further stipulate that, in the absence of a cogent alternative like evolution, the analogy with organisms (that they, too, bespeak a knowledgeable, conscious intelligence behind them) was a plausible argument—for the 1820s. But how compelling is the analogy today? The argument boils down simply to this: we can invoke a naturalistic process, evolution, for which there is a great deal of evidence, but which we still have some difficulties in fully comprehending. Or we can say, simply, that some Creator did it and we are, after all, only complex machines like watches. The analogy is as meaningless as that: it proves nothing. *It could even be true*, but it cannot be construed as science, it isn't biology, and in the end it amounts to nothing more than a simple assertion that naturalistic processes automatically cannot be considered as candidates for an explanation of the order and complexity that we all agree we do see in biological nature.

To bolster the argument from design, creationists jump to the other side of the complexity argument: evolution just could not, they say, produce these organic complexities, because there is no way such complex structural systems could have developed by chance alone. Just as a bunch of monkeys endlessly pounding typewriters would never duplicate the works of

Shakespeare, they argue, no mindless, materialistic process such as evolution—portrayed as acting by blind chance alone—could ever have produced the myriad wonders of the organic realm.

Evolutionary theorists are not the simpletons such statements would make them out to be. As we have seen, evolutionary biology has been very clear on just this point: mutations are random, but random only with respect to the needs of an organism. Mutations, insofar as most geneticists are concerned, do not arise because they might be useful to an organism. On the other hand, mutations are caused by real physico-chemical processes, and there is a limited number of forms that a mutation can take and still function as a viable gene. In this latter sense, mutations are not random: there are a limited number of biochemical changes that a gene can undergo.

The antichance element of evolution is, of course, natural selection. Richard Dawkins has ably described this crucial deterministic aspect of evolution in his book *The Blind Watchmaker* (1986).[14] And though biologists have with some justification referred to natural selection as the "creative" force in evolution—governing, as it patently does, the development of novel structures, behaviors, physiologies, and biochemical pathways—natural selection really is just a stolid bookkeeper, i.e., *not* a watchmaker. In a world of finite resources, on the whole it is the organisms best suited to making a living that will survive long enough to reproduce, and it is their genetically based properties that will differentially be passed along. With each generation, genetic recombination presents new packages of "variation"—the ultimate source of which is mutation—to the environment, and that is what determines—in a statistical manner—what will be passed along to the next generation. It is silly to think of natural selection as somehow the equivalent of the creationist's Creator-God. Rather—and far less grandiosely—natural selection is the natural antichance process governing the transmission of genetic information within populations from one generation to the next. In other words, natural selection is the answer to the creationist statement that the diversity of life cannot have been produced by chance alone, and natural selection should not, in any sense, be given godlike status.

So chance, design, and complexity are handled adequately, if not always stunningly, by evolutionary theory and in biological observation and experimentation—sufficiently well to be scientific on the one hand, and not to require the ad hoc intervention of a supernatural Creator on the other.

## The Origin of Life

What about the origin of life? For creationists, the origin of complex, self-replicating living systems from the inorganic realm demands the action of improbable chance and implies a Creator. Pointing to the inability of biochemists to synthesize life in a test tube, creationists agree with the poet: only God can make a tree. Only a Creator could have assembled all those complex ingredients of DNA, house them in a proteinaceous sheath, and thus fashion the first primitive form of life.

Evolutionists commonly respond that complex organic molecules occur throughout the universe and many (such as amino acids, the building blocks of proteins) can be synthesized simply by passing a spark through a gaseous mixture of ammonia, methane, hydrogen, and water, as was first done by Stanley Miller in the 1950s using the ingredients thought to be the main atmospheric constituents of the primitive Earth. Creationists counter that such results are far removed from producing true life. Biologists, of course, agree, while maintaining that such experiments are both supportive and suggestive of the hypothesis that life did, indeed, arise from natural processes.

Some biologists, such as Nobel laureate Francis Crick, do stress the great difficulties involved in the origin of, say, the molecules of inheritance and protein synthesis—DNA and RNA—from simpler, and ultimately inorganic, systems. Such biologists seriously doubt the ubiquity of life throughout the universe as envisioned by some cosmologists, who argue that an improbable event becomes probable given enough tries: there are billions of stars in the universe and so, one may suppose, many planets with conditions similar to our own on which life may well have developed independently. Neither argument is particularly compelling in the absence of any hard information. But it is important to note in passing that whether or not life arose on Earth, or arose elsewhere and spread here (a view favored by Crick, for example), both sides agree that once bacteria became established on Earth, all the rest of life, as we know it, evolved from them.

It is true that DNA is complex. It is true that no one has taken primordial compounds supposedly in the Earth's primitive atmosphere and created DNA—much less a functional bacterium—in the laboratory. The creationists wish us to suppose that this situation demonstrates that life cannot have arisen by natural processes. I cannot follow their argument: in the brief history of biochemistry we have gone from laborious analysis of what proteins are (starting in the mid-nineteenth century), through the cracking of the genetic code (in the mid-twentieth century), to the heady days of genetic cloning (at the end of the twentieth century). That the origin of life, if posed as a biochemical problem, remains incompletely solved as of the year 2000 is not particularly surprising and certainly not compelling evidence that it never will be. But if we are to continue to teach our children that such problems are beyond the purview of science because "the Creator did it," we certainly will lessen our chances of ever finding out. Yet that's exactly what creationists—including Phillip Johnson and his colleagues—would have us all think.

## The Last Word: A Simple Refutation of Creationism

*There must be a single, hierarchically arranged pattern of resemblance interlinking all life if all life descended from a single common ancestor.* This is evolution's grand prediction, and as we have seen, it has been abundantly and consistently corroborated throughout the annals of biological research.

What do creationists offer as their explanation for the manifestly hierarchical structure of the biological world? Most creationists simply affirm that it pleased the Creator to fashion life in the form in which we find it today. They maintain that the Creator was simply being efficient in using the same blueprint for the separately created basic kinds, thereby "explaining," for example, why mammals, birds, reptiles, and amphibians all have one upper and two lower leg bones (except, of course, snakes and other secondarily limbless tetrapods).

There is a simple test of the proposition that the hierarchically arranged structure of life is the product of intelligent design—of a Creator-God. It comes from the very same watchmaker analogy that creationists apply in their arguments on complexity, for if it is true that no one can devise a direct way to observe the behavior of the Creator, we certainly can do the next best thing: we can examine the history of watches—or any other product that humans have designed over time. And we ask, Does the design history of watches—or 747s, or automobiles, and so on—reveal a simple, hierarchically nested pattern of similarity, as evolution has produced in the biological realm? The answer is a resounding *no*.

By sheer coincidence, I happen to be an expert in the history of design of the cornet, the brass musical instrument that is the shorter, dumpier version of the more familiar trumpet.[15] Cornets were invented in the 1820s, and the basic configuration of modern cornets was established by the mid-1850s. For the past century and a half, a bewildering array of cornet designs has appeared—an exuberant variety that has defied all attempts at neat categorization. I tried for several years to produce a simple classification of cornets—one that resembles the classifications I have produced for trilobites and horseshoe crabs in my career as a paleontologist. And I persistently failed.

I think I know why human design systems can never yield the same sort of simple patterns that we see in the biological world. The reason is that humans are continually copying each other—and stealing each other's ideas. It was one thing for monks to copy manuscripts, which would then be copied again *in isolation* in far-flung monasteries. Copying in isolation does, as we saw in Chapter 2, produce the same simple hierarchical structure that we observe in the natural world. But out in the competitive marketplace, it is another thing altogether, and cornets of every conceivable blend of design have been produced. The result is a mélange of design that defies simple characterization and unambiguous classification. On the face of it, when we examine the only examples of intelligent design open to us, we see that the prediction that intelligent design would produce the same sort of simple hierarchically nested pattern that we observe in the biological realm fails utterly.

However, the Creator is not supposed to have had competitors in designing the biological world. The real analogy would be with the output of a single artisan or atelier. I have chronicled in detail the design history of two major cornet manufacturers, and I am thoroughly conversant with the design histories of at least a dozen others. Though there is a tendency for makers producing more than one design to keep them separate for years on end, there are plenty of examples of the blending of designs through time. And I know of only one case in which progressive modification of a design lineage in several discrete steps yields a simple, hierarchical classification scheme.

Could the single artisan, who has no one but himself from whom to steal designs, possibly be the explanation for why the Creator fashioned life in a hierarchical fashion—why, for example, reptiles, amphibians, mammals, and birds all share the same limb structure? Here, and somewhat to my surprise, I find myself agreeing with creationist Gary Parker, who says that God did not work that way at all. Denying that life really is hierarchically structured, Parker writes (alluding to patterns of similarity in different organ systems in lizards, but clearly generalizing), "The pattern is not a branching one suggesting evolutionary descent from a common ancestor; rather, it is a mosaic pattern . . . suggesting creation according to a common plan" (1980, p. 22). I would agree with Parker: if I were designing life, I probably would use the same idea over and over—but not, as Parker suggests, limiting the use of my ideas to what someone in retrospect might be tempted to identify as separate lineages. Good design ideas, in other words, should turn up here and there all over the place, wherever they prove useful, and whether or not they are stolen from others or are the product of a single fertile mind.

So, in the end, there is as little of substance in the scientific creationists' treatment of the origin and diversification of life as there is in their treatment of cosmological time. They pose no novel testable hypotheses and make no predictions or observations worthy of the name. They devote the vast bulk of their ponderous efforts to attacking orthodox science in the mistaken and utterly fallacious belief that in discrediting science (or, as they put it, evolution science or philosophical naturalism), they have thereby established the truth of their own position.

Their efforts along these particular lines are puny. Moreover, they impugn the integrity and intelligence of thousands of honest souls who have had the temerity to believe that it is both fitting and proper to try to understand the universe, the Earth, and all its life in naturalistic terms, using only the evidence of our senses to evaluate how truthful an idea might be. Yes, historical geology and evolutionary biology are sciences. They are imperfect—but self-correcting. And no, neither scientific creationism nor Johnson's theistic realism can be construed as science—not by any conceivable stretching of the term. And if it is not science, what is it? Phillip Johnson has already let the cat out of the bag: creationism, including special creationism, scientific creationism, and theistic realism, is nothing but that good, old-time religion.

# CHAPTER 7

# Can We Afford a Culture War?

Phillip Johnson thinks that what is at stake here, in his words, is a "culture war" between the atheistic forces of naturalism on the one hand, and an essentially Christian-based (his version!) view of the origin and nature of things—with all the moral, ethical, and, yes, legal and political implications these would seem to imply to his basically conservative viewpoint—on the other. On the campus, the alternative to philosophical naturalism, Johnson fervently hopes, is to be his theistic realism, but far more is at stake in the body politic at large.

Some of my colleagues patently agree. In debate, Will Provine says that people have to "check their brains at the door" when they go to church. Others are all too willing to agree that Darwinism implies atheism—blindly rising to the bait, dismissing as "nineteenth-century fairytales" the sort of personal God espoused by Johnson (and so many others, of course), and grossly overstepping the limits of what can be said from the actual point of view of science. Scientists like Richard Dawkins seem only too willing to agree that we are indeed involved in a culture war.

I see this war, but I think it is both overstated and very risky. I do admit that there is a huge political side to it; indeed, I think that is really where the war is. Academics—Johnson and Provine alike—like to think that everything is about ideas (actually Johnson, a lawyer, knows better), where in contrast it should be fairly obvious that the essence of all this creationism fury, despite its deep-seated roots in one particular branch of religion, is not so much good old-time religion, but good old-time politics. I myself am politically left, though as a middle-class suburbanite who enjoys the niceties of life to the extent I can afford them, I am hardly a radical. I have already said that I harbor extreme doubts on the existence of such a personal God as the one in which Phillip Johnson believes (though I'll stick to my promise and say more about this in the final section of this chapter). Unlike Will Provine, though, I do not see it as my business, nor important in any intellectual sense, to attack anyone's religious beliefs—or to worry what anyone personally might think about evolution or the biblical stories of creation.

But I do think the issue of creationism in public schools is *very* important—enough so to warrant writing this book. Though intelligent-design arguments seem to be on the ascendancy as the strategy of choice when it comes to combating evolution in the public schools at the end of the twentieth century, we still are not entirely done with the older form of scientific creationism, which seeks to establish creationism as a bona fide form of science and thereby teachable in public schools in America. The real battle is still being fought at school board meetings and in public school classrooms, where local creation enthusiasts (sometimes, but not usually, in cahoots with a creationism-leaning teacher) persist in trying to inject their version of Christian theology into the public schools. Nor is Phillip Johnson alone in pretending that the tired old arguments against evolution are brand new: every year, new efforts to debunk evolution and establish the teaching of creationism—either instead, or at least alongside, of evolution—trumpet the "news" that evolution has been falsified. Creationist literature reads like the worst of the supermarket tabloids—the ones that tell you there are Martians after all, though somehow you missed the news in the *New York Times* or on CNN.

According to the National Center for Science Education—an anticreationist, pro-science organization located in Berkeley, California, that grew out of early "Committees of Correspondence" that were organized in the 1970s and early 1980s on a state-by-state basis—the teaching of evolution in the late 1990s is in as much serious trouble in the public schools of the United States as it has ever been. Alabama actually requires a disclaimer in its high school biology books that reads, "Evolution is a controversial theory some scientists present as scientific explanation for the origin of living things, such as plants, animals, and humans. No one was present when life first appeared on earth. Therefore, any statement about life's origins should be considered as theory, not fact."[1] Seven other states are said to avoid evolution as much as possible in the biology curriculum, and no fewer than 25 states have had recent difficulties fending off creationism within isolated local school districts.

So we must ask again, Why does the problem persist? And why should we care about it?

I firmly believe that the world is the way it is regardless of what anyone thinks about it—you, I, the president of the United States, the kid down the block. I am aware of the sophisticated arguments in the philosophy of science that, because any statement we make about the natural world is necessarily a mental construct, there can be no wholly objective reality—and I agree, at least to the extent that nothing we say about the material world can be counted on as absolute. On the other hand, I remain enough of a logical positivist to maintain that there is a physical reality, and we are not merely constructing it when we look critically at the stars, subatomic particles, or Devonian trilobites. It is not, in other words, a purely unobjective—subjective—statement to say that the Devonian trilobites on which I work were arthropods living in seas some 480 million years ago. On this, I'm with that other famous Johnson—Samuel (1709–1784)—who, when commenting on Bishop George Berkeley's (1685–1753) claim that the physical world is an illusion created by humans, is reported to have kicked a rock while saying, "I refute it thus." In other words, I don't think that when I walk out a door, everything in a room—the furniture, the people—cease to exist.

That is what I mean when I say that the world is the way it is regardless of what anyone says it is. Someone—say, Phillip Johnson—can swear that evolution has not happened and have absolutely no effect on the relationship between humans and chimpanzees. On the other hand, I can express my extreme doubt that the sort of personal God that Johnson knows is there is actually taking a direct hand in my affairs—or those of anyone else—and that should have no effect whatever on anyone else's personal construal of

God, or, for that matter, how they think the universe ticks—or who puts the bike under the Christmas tree.

All that being said, though, it very much does matter what we teach our kids at school. In the biology classroom, it simply cannot be that "my opinion is as good as yours"—not if we are to teach with integrity in the science curriculum. Anticreationists like me have for years agreed that "origin" accounts—of the Judeo-Christian tradition, but also of the Buddhist, Confucianist, Hindu, and other major religious traditions, as well as of different Native American, African, and Asian traditions, not forgetting hunter-gatherer traditions as recorded by anthropologists—are all grist for the mill of a comparative course on religious beliefs.

The scary thing about this is not that a kid might prefer to believe the creation story. After all, some 44 percent of Americans, according to one poll I saw recently, do adopt this position. The purpose of teaching science is not to indoctrinate kids on the (secular) humanist or naturalist side of the culture war in which we are supposedly engaged, but rather to teach them what science is all about. No good teacher will demand that kids "believe" evolution; kids should never be taught anything other than "this is what science thinks about this issue"—evolution, plate tectonics, quarks, and so on—and this is *why* scientists think this. In short, they should get a very clear idea about how science is done—how it *works*—and especially come to see science as a perfectly human enterprise (with all the implicit failings of anything human!). *Especially* when kids come to class espousing creationism and showing resistance to hearing anything about evolution, the teacher must work hard to make it clear that the kids' beliefs will be respected, and that they won't be asked to drop their religious beliefs and adopt a new belief in evolution.

Belief simply is not the point here; rather, a thorough grounding in science as a human endeavor is. This is important simply because we live in a technological age so heavily dependent on science—with no realistic thought or hope of turning back—that the future of this country, of the Western world generally, and undoubtedly of humanity on the planet as a whole depends very much on more and better science, especially on an informed citizenry who must continue to guide the future course of this and all other countries as wisely as possible through the ballot box. I would no sooner place our future strictly in the hands of scientists than I would see it placed in the hands of movie stars—or, for that matter, lawyers and politicians. I think the problems facing humanity at the Millennium are so great that we need the input of all segments of society to deal with them, and here I refer specifically to perhaps the greatest sector of society to which one can point: the global community of organized religion.

We will not go very far if we pretend to our kids that we cannot tell the difference between real and phony science. Yet that was the gist of all those "equal time" laws of the 1970s and 1980s: the Arkansas and Louisiana legislatures were actually telling the teachers in their public schools to pretend not to know the difference between real science—flaws and all—and outmoded or simply bad science. I cannot imagine anything more perverse, more deliberately harmful, than teaching kids that their elders cannot tell the difference between the real and the phony. Some of them, of course, cannot. But all but the relatively few creation-leaning science teachers throughout the fifty states most assuredly can, and requiring them in essence to lie to their students sends about the worst message imaginable to the younger generation. And kids, of course, can see right through that.

That's why I so fervently care about teaching evolution, and not teaching creationism, in public schools. It is not that I want kids to abandon their religious beliefs; it is that I want our kids to be able to know science for what it is, so that they can make informed choices as adults. I want people neither to follow science slavishly as if it were the only salvation—the only way of knowing—nor to condemn it outright for all the evils (real and imagined) that it has unleashed on the world. Science is a "glorious enterprise," and kids simply have to learn about it so that they may see it for what it is.

Why, then, this particular animus against evolution? Why has it persisted so long in the American political arena? Creationists attack only the part of science that they find inimical to their religious beliefs. Moreover, though the attack on evolution comes largely from the Christian right, not all political conservatives are Christians, and not all Christians, even conservative Christians, are political conservatives. Nor is it just a matter of "my story versus yours," though early opposition from Christian fundamentalists, who simply thought that anything that cast doubt on the literal truth of any part of the Bible (biblical inerrancy) cast doubt on the whole of Christian doctrine, did indeed amount to "my story against yours." And though it is also true that, when a Phillip Johnson debates a Will Provine, they still agree that it is "my story against yours" (with each insisting he has the truth, of course), it is just as obvious that something more is at stake than simply God's credibility as divined from the pages of the King James Version of the Bible versus modern biological research.

That something is morality. Perceived decline in moral values in the United States and perceptions of what might be done about it are what prompt the political right's continual war on evolution. But it is by no means just the American political right that sees a connection between evolution and what philosophically inclined biologists (and biologically inclined philosophers, such as Michael Ruse) prefer to call ethics. Starting with Herbert Spencer's social Darwinism, the tradition to develop ethical systems and, sometimes, paths of overt policy (e.g., as in the eugenics movement[2]) based on biological principles has persisted in Western culture.

In a sense, the Christian right's outright opposition to evolution is just one aspect of this hypnotic temptation to see moral or ethical implications in evolutionary biology. Phillip Johnson is glad to pounce on biologists and philosophers who have dabbled in these waters—as evidence that there is indeed a culture war taking place. I am happy to report, though, that in my lifetime few ethical ruminations derived from evolutionary biology have made it all the way to the body politic. And that is a source of some relief, for what a confusing welter of ethical systems and homilies have been drawn up in the name of evolution! I have seen very similar ethics derived from diametrically opposed evolutionary camps and, of course, very different ethics derived from essentially the same camp within the larger evolutionary circle. For example, the great geneticist Theodosius Dobzhansky, a devout and practicing Christian, derived what I can only describe as a gentle, loving worldview of, among other things, cooperation—believing deeply, as he did, that Christianity and evolution went hand in hand.

Were he alive today, I am reasonably sure that Dobzhansky would not identify with the strongest, most strident gene-competition views that underlie sociobiology. Yet natural selection is the quintessential cornerstone of all evolutionary theory in general, and surely of the genetics of both Dobzhansky and, say, Richard Dawkins. And it is notorious that the "selfish gene" underlying sociobiology has conjured up an ethical worldview to some contemporary biologists and philosophers that is very like the dog-eat-dog vision of the worst of the old social Darwinism.

Well, if evolution can prompt ethical systems of ruthless competition in some minds, and Christian-like harmony in others, what are we to conclude? Here is what I have long thought: *there is no one-to-one correlation between any principle of science and any system of human behavior. In particular, there is no necessary set of ethical implications implicit in the very idea of evolution—or emanating from any subset of evolutionary theory.* To those who say there are moral lessons and ethical systems—evil or good—implicit in the very idea of evolution, I say, A PLAGUE ON BOTH YOUR HOUSES.

There is no doubt that the creationists' tilt against what they wish was the evolutionary windmill is born in greatest part by the sincere belief that there is indeed a connection between evolution and morals—and a negative one at that. Just since I began dueling with creationists in the late 1970s, I have either experienced, or seen secondhand, a number of events that dramatize how deeply the connection between evolution and moral decay is seen by the Christian right. For example, during the "Scopes II" trial in California, Nell Segraves (the plaintiff's mother and a director of the Creation Research Center in San Diego) had the following exchange with Robert Bazell, science reporter for *NBC Nightly News:*

Bazell: "For seventeen years since the Supreme Court banned prayer in public schools, Mrs. Segraves has been fighting to bring religion back to the schools. She believes that the teaching of evolution is the primary evil, responsible for all sorts of problems."

Segraves: "What about prostitution, or the drugs, or the criminal activities, violence. It's lack of respect."

Bazell: "And you think that all that can be traced to the teaching of evolution in the schools?"

Segraves: "I believe it can, and I think I can prove it."

Now it is surely an irony that the Old Testament amply documents the presence of many of the same social ills (and plenty more) plaguing Israelite society thousands of years ago, yet nothing is said of the Israelites teaching evolution to their children.

Creationist R. L. Wysong is quite explicit on the reason why evolution leads to moral decay. In his *The Creation-Evolution Controversy* (1976), Wysong argues that one's position on origins frames one's worldview, and one's worldview in turn leads to one's "approach to life." A "correct" position on origins leads to a "correct" worldview, which in turn leads to solutions to life's problems.

> A person can basically take one of two positions on origins. One is there is a creator, the other is there is not; or, evolution explains origins or it does not, Creation versus evolution, theism versus materialism or naturalism, and design versus chance, are all ways of expressing the two alternatives. (pp. 5–6)

Wysong goes on to argue that the evolutionary position influenced the thinking of such historical figures as Marx, Mussolini, Hitler, and Freud:

> but if, on the other hand, life owes its existence to a creator, a supernatural force, then life is the result of his will and purposes. Understanding these purposes would be the only way to understand life's varied questions and problems. (p. 9)

My favorite example comes from a woman in Philadelphia, who, in a letter to me written in the early 1980s, put it even more bluntly. Greatly mistaking my position (taken from a newspaper article), she wrote to thank me for my efforts in combating the notion that we humans have descended from lower forms of life, for *were we to teach that to our children, we could not expect them to conduct themselves in a moral way*. In so writing, she cut to the very heart of creationist feelings.

One need not be a materialist to note that a functioning society demands moral behavior, and that there are other, more compelling, explanations for why we are, collectively, less than perfect in our ability always to behave in the very best way. But such is the creationist belief, and the reason for their persistence is, far more than the urge for fame or the slight profit motive attributable at least to some creationists, their conviction that evolution undermines morality.

More recently, the same rhetoric has been popping up in pronouncements from right-wing political figures. Conservative columnist and erstwhile (perennial?) presidential candidate Pat Buchanan, for example, has expressed his belief that the American public has "a right to insist that Godless evolution not to be taught to their children or their children not be indoctrinated in it."[3] And, taking us right up to within a few weeks of this writing, according to a New York Times editorial, House Republican Whip Tom DeLay has also joined in:

> In the culture wars, you can credit Mr. DeLay for turning a major political axiom upside down. It used to be an article of faith for conservatives that Americans need to take more individual responsibility for their actions. But now, thanks to Mr. DeLay, we learn that violence perpetrated by gun owners is really the product of larger forces. What might those be? According to the Republican whip from Texas, nothing less than "broken homes," day care, television, video games, birth control, abortion, and, *unbelievably, the teaching of the theory of evolution* [italics mine].[4]

The most visible battles of this particular culture war are the cases that have actually found their way to the courtroom from time to time, starting with the notorious Scopes trial of 1925. Scopes was found guilty of violating the Tennessee statute known as the Butler Act, which said in part,

> It shall be unlawful for any teacher in any of the Universities, Normals and all other public schools of the State, which are supported in whole or in part by the public school funds of the State, to teach any theory that denies the story of the Divine Creation of man as taught in the Bible, and to teach instead that man has descended from a lower order of animals.

Clarence Darrow and his colleagues, including a battery of lawyers from the American Civil Liberties Union, never disputed the charge that John Scopes had indeed taught the evolution segment of the biology curriculum. (It is an amusing sidelight that, later in life, Scopes admitted he had in fact never taught evolution to his biology class—though he had covered the subject in general science.) Convicted of his "crime," Scopes was fined a hundred dollars. The Tennessee Supreme Court then threw out the conviction on the technicality that Judge Ralston had improperly levied the fine on Scopes because Tennessee law mandated that only a jury could impose fines of fifty dollars or more.

The action stymied the plans of Darrow and colleagues: their real aim was to have the law reviewed and thrown out by the U.S. Supreme Court. Their argument, of course, was to be that the law stood in violation of the Constitution, particularly the establishment clause of the First Amendment prohibiting the mixing of church and state.

The issue of creationism in the classroom finally reached the U.S. Supreme Court in 1968 in *Epperson v. Arkansas*. Mrs. Epperson, a Little Rock high school biology teacher, successfully challenged a 1929 Arkansas law forbidding the teaching of "the theory or doctrine that mankind ascended or descended from a lower order of animals." Not until the Supreme Court ruled that the Arkansas law was "an attempt to blot out a particular theory because of its supposed conflict with the Biblical account," and thus was an attempt to establish religion in the classroom, were the Tennessee and other similar statutes declared null and void.

Certainly not legally dead as a result of the Scopes trial, creationism maintained a steady though low profile through the next fifty years. I maintain my conviction that the real battleground is the classroom, and it is here that we have seen the most subtle yet profound sign that the creationists, in an important sense, actually won the day in 1925, for thereafter there was a dramatic downplay of evolution in high school biology texts from the late 1920s on. A case in point: George W. Hunter's *Civic Biology*, published in 1914— the book Scopes testified he had used in his course—originally had a brief discussion of evolution. After the trial, the publishers brought out a revised edition *(New Civic Biology)* eliminating any mention of evolution. Virtually all other texts followed suit. If Genesis had not quite made it into the classroom, the Scopes trial did at least place a tremendous damper on the teaching of evolution for the next thirty-five years or so. Only when Americans awoke one day in 1957 to see Sputnik circling the Earth—and awoke thereby to the deficiencies of science education in the United States—was anything done. The resulting massive, national effort to upgrade the quality of science education included evolution, in a sense priming the pump for the battles still under way in many of our secondary schools.

Then in the 1960s, Henry Morris and his colleagues invented scientific creationism, the wolf in sheep's clothing devised for the very purpose of circumventing the establishment clause of the First Amendment—the one that the Supreme Court invoked in 1968 when it struck down those archaic old "monkey laws." Though similar laws were pending in the legislatures of many states in the 1980s, it was Arkansas Act 590, followed soon thereafter by passage of a similar statute in Louisiana, that triggered the two famous creationism court cases of the modern era. These were "equal time" bills, which, as we saw in the case of the Arkansas statute, called for the teaching of a model of creation science alongside of evolution science. As we saw at the outset of Chapter 5, the definition of creation science in Arkansas Act 590 is virtually identical to lawyer Wendell Bird's creation science model, already encountered in Chapter 5.[5]

I was briefly involved in the preparation of the plaintiff's case against Arkansas Act 590 that was successfully argued in Judge William Overton's Little Rock courtroom in 1981. I ended up attending the trial, though, as a reporter, armed with my very own press pass. But rather than conjuring up old memories, I have chosen to reprint as Appendix 1 the "review" of the trial I wrote for my sponsoring publication, Science 82, a now defunct publication, aimed at the general public, of the American Association for the Advancement of Science, which publishes the very much still alive journal Science.[6]

What really is important, of course, is Judge Overton's opinion—widely considered to be so tightly reasoned and documented that there was little room for appeal. Overton found that Arkansas Act 590 violated all three of the by then traditional litmus tests of whether or not a statute is in violation of the establishment clause of the First Amendment to the U.S.

Constitution, pertaining to the relation of church and state. Roger Lewin's report of Overton's opinion (see note 6), provides the following citation from a 1971 opinion that established the three-pronged "Lemon test" of the establishment clause:

> For a statute to be constitutional, it must fulfill three provisions: "First, the statute must have a secular legislative purpose; second, its principal or primary purpose must be one that neither advances nor inhibits religion . . . ; finally, the statute must not foster an excessive governmental entanglement with religion."[7]

Violation of any one of these three criteria would have been sufficient to rule Act 590 unconstitutional, and Overton saw violation of all three!

The strategic mistake the Arkansas legislature made when drafting Act 590 was to include explicit definitions of evolution science and creation science. The expert witnesses especially tore through the fatuous creation science characterizations, and evidence was also adduced demonstrating the religious motivations of the bill's original sponsor.

Less clear-cut was the Louisiana case, argued in 1982: *Edwards v. Aguillard.* Here there was no explicit definition of creation science—only a much vaguer statement saying that "creation science means the scientific evidences for creation and inferences from those scientific evidences." No explicit definition, no prominent fish swimming in the barrel for expert witnesses to shoot at. But the Louisiana court, nonetheless, did find the statute unconstitutional. The case eventually made it all the way to the U.S. Supreme Court, which ruled, 7 to 2, that the statute is indeed unconstitutional on "establishment" grounds. Only Justice Antonin Scalia and Chief Justice William Rehnquist demurred, finding the Louisiana law in fact constitutional.

Though I have drawn attention to the disproportionate number of lawyers (e.g., Wendell Bird, Norman Macbeth, Phillip Johnson) who have come to the forefront among the layperson Darwin attackers and outright creationists, I must say that I took deep satisfaction, sitting in Judge Overton's courtroom all those years ago, in seeing that the argument that "creationism is really science, so we can teach it in the classroom" is as transparently pre-posterous to the interpreters of the law of the land as it is to anyone else who is not a creationist zealot. At least in this instance, the law proved itself *not* the "ass" that Charles Dickens's character Mr. Bumble (in *Oliver Twist*) once pronounced it to be.

Thus the legalistic charade culminating in the 1980s has—at least for the nonce—faded, and we are left with what the likes of Luther Sunderland was on record as advocating: not the conspicuous activities of legislatures and the inevitable court cases, but the much more insidious, quiet, and, sadly, effective pressuring of school boards and teachers at the grassroots level. And that is still going on. It is a shame, really, because there are some press-ing problems that present a much brighter way to look at biology and reli-gion within the larger body politic. To see this connection, we must be willing to suspend the by now hoary, wholly outmoded fight over origins, and be willing to take a fresh look at the world around us. Never mind the origins of biological diversity, including human life. Rather, we simply acknowledge that life is here, that it is in trouble, and that nonhuman life remains important to human life. When we do that, we get a whole new perspective of the connections between science and religion—*positive* con-nections—that help point the way to solving some of the most critical prob-lems faced by life on this planet as we enter the new Millennium.

## Ground Zero

There are some 10 million species (a conservative estimate) living right now on Earth. Of these 10 million species, we are losing roughly thirty thousand a year (again a conservative estimate).[8] Scientists so far have discovered and named just under 2 million of these species, but we know there must be many more, since random handfuls of backyard dirt reveal countless microbes and nematode worms as yet undescribed, and virtually every tree felled in the Amazon rainforest has an insect never before seen. But we *know* we are losing lots of these species as ecosystems are systematically destroyed. Humans, for example, cut down many hectares of rainforest a day, and given the known ranges of tropical species, we know that some-thing like three species an hour are disappearing—most of them before we even have a chance to call their names. We are, in short, in the midst of an extinction crisis of proportions the Earth has not seen since the last of the five major, global mass extinctions struck the Earth some 65 million years ago.

The question becomes, Should we care? After all, human life seems to go along fine in New York, no matter how many trees are felled in the Amazon rainforest. But when we stop to consider that human beings around the planet utilize over forty thousand species of plants, animals, fungi, and microbes every day—for food, shelter, clothing, and medicine—we get an inkling of how important nonhuman life on the planet still is to human existence. We need our fisheries and forests; we need our genetic reservoirs for the production of new medicines, and for the genetic replenishment of existing agricultural stocks.

We recognize that the purely physical aspect of human existence—just being alive, breathing the air, drinking the water, converting plant and animal proteins, sugars, and starches for the energy and proteins we need to maintain, grow, and simply stay alive—depends on the availability of these precious commodities. Right now there are just over 6 billion human beings on Earth, and only roughly half of them have access to fresh water safe enough to drink. Topsoil is being stripped from the centers of the major continents at such high rates that the ever greater amounts of food we need to produce to fill the ever increasing number of human mouths is also at great risk.

It is the onslaught against ecosystems themselves that (just as in the past) accounts for species loss, but also for the degradation in air quality and water supplies, and the cycling of essential nutrients like carbon, nitrogen, and phosphorus. In short, we humans are fouling our nest at such a rapid clip that, I think it is safe to say, our own continued existence is called into question. We need to stem the tide of ecosystem destruction and species loss if we are to hope for a long and fruitful continued existence on this Earth—at least at the levels of material comfort and "high civilization" that the industrialized world currently enjoys.[9]

So much for the problem—the issue at hand. What can this possibly have to do with religion?

Everything. Over the past few years I have visited a dozen or so college campuses, discussing aspects of "science and religion" with my colleague Margaret Wertheim. I have also spoken at various "ecology and religion" symposia, and even at the United Nations (fall 1998) on this subject. What I have to say at these venues is simple: I am convinced that, to understand how humans have come to be destroying ecosystems and driving species to extinction at such prodigious rates and intensities—an understanding that is necessary to define strategies to overcome the problem—we must understand the fundamentals of the human ecological condition. And I present my scientific analysis of three distinct aspects of the human ecological condition over the past ten thousand years (i.e., within the time frame adopted by the young-Earth creationists). But I also say this: *religious traditions, especially as embodied in concepts of God, are deeply if not wholly ecological concepts as well.* And I tell my audience that there is, in effect, a resonance—not disagreement—between the modern scientific account and traditional religious accounts. Not surprisingly, people have always had a pretty shrewd grasp—not necessarily of where they came from in the first place, but of who they are and how they fit into the world around them. And they tend to reflect these deep understandings, their visions of themselves and how they fit into the world around them, in their religious accounts, including the concept of the gods, or of God.

Here, very briefly, is what I say as I compare these three parallel accounts:

Scientific Analysis I. The "state of nature" of absolutely every species that has ever lived on the planet—including, until very recently, all humans—is to be broken up into relatively small local populations living as a collective part of a local ecosystem. Each such ecosystem has a carrying capacity, a limit to the population size that is determined by the amount of food productivity, water accessibility, and other aspects of the environment, according to the behaviors and physiologies (evolutionists like me would say "adaptations") of whatever species one is considering. The role each such local population plays within its ecosystem is what is generally meant by ecologists as the ecological niche of that population.

This description of local populations of species having distinct niches within local ecosystems does not apply to the vast majority of humans now alive. However, some remnant hunter-gatherer cultures will survive into the twenty-first century—if just barely, but with their traditional cultures in all cases already radically altered. Fortunately, we have abundant documentation from anthropologists (and others, including missionaries) who studied many of these cultures in the relatively early days of Western exploration and colonization. We know that, traditionally, these peoples lived in relatively small bands and relied exclusively on the productivity of their local surroundings, meaning that they conformed exactly to the description of ecological niche occupation that is true for all other species on the planet.

Concepts of God I. Concepts of supernatural beings among hunter-gatherer peoples reveal a clear and starkly direct understanding of who they are and how they fit into the world—in effect, the universe—around them. One example I give is reported by anthropologist Colin Turnbull, who studied the Ba Mbuti people (so-called pygmies) of the Ituri forest in Africa's Congo Basin.[10] Turnbull tells of a group of Ba Mbuti—men, women, and children—venturing into the forest on a joint hunting and plant-gathering foray. As they entered the forest, they called out, "Hello Mother Forest! Hello Father Forest!" And they would tell Mother and Father Forest that they were there to take only what they needed.

The Mother and Father Forest of the Mbuti are gods—spirits of the natural world. The Mbuti are, in effect, clearing their hunting and foraging with these spirits of the forests, and by insisting that they are there only for what they need, they are acknowledging in the clearest possible terms that they see themselves as a part of the very forest ecosystem in which they live. The spirits they invoke—Mother and Father Forest—are *abstractions* of the system in which they live.

Resonance—I have no other word for it: my Western biological understanding of what constitutes an ecological niche and how species are broken up into small populations that form parts of local ecosystems—a description I claim pertains to hunter-gatherers—is mirrored precisely in the accounts the hunter-gatherers themselves provide, and it is especially clearly seen in their concepts of the supernatural.

Scientific Analysis II. Sometime around ten thousand years ago, in a variety of places around the globe—but perhaps first in the Natufian culture of the Middle East—people began actively to cultivate crops and undertake animal domestication in a systematic fashion. Though often described as an expansion of the human ecological niche, it was really far more momentous, for *with the invention of agriculture, people radically changed their position in the natural world, becoming the first species in effect to "step outside" local ecosystems.* If anything, the local ecosystem became the enemy—with all but one or two of the local plants now regarded as invasive weeds, and all but a handful of the local animals viewed more for the harm they could do to crops and livestock than as a part of the natural system.

In short, the invention of agriculture meant that *Homo sapiens* became the first species ever to exist on Earth *not* to be broken up into local populations inside of local ecosystems. In these brief ten thousand years, the human population has grown from an estimated 6 million to over 6 billion. There is no question that this surge in population (it has been logarithmic, with the big increases, of course, coming just in the last one and a half centuries), plus the unequal distribution and consumption of resources, is what underlies the conversion of ecosystems to farmlands, and later to cities and suburban settings—the prime cause, in other words, underlying the extinction crisis we now face. It is the very success of humans in taking the bold step out of ecosystems that has led to the current ecological predicament.

Concepts of God II. Did the early agriculturalists have as fine-tuned a sense of who they were and how they fit into the world around them as the Ba Mbuti? There is no doubt in my mind that they did, and I can cite no better source than my old King James Version of the Bible. I have, in the past,[11] quoted Genesis 1:26–28—the famous "dominion" passage—in which God is said to have said,

> Let us make man in our image, after our likeness: and let them have dominion over the fish of the sea, and over the fowl of the air, and over the cattle, and over all of the earth, and over every creeping thing that creepeth upon the earth.

So God created man in his *own* image, in the image of God created he him; male and female created he them.

And God blessed them, and God said unto them, Be fruitful, and multiply, and replenish the earth, and subdue it, and have dominion over the fish of the sea, and over the fowl of the air, and over every living thing that moveth upon the earth.

These words constitute the most ringing declaration of independence ever set down. They say that people, whatever their similarities with the beasts of the field, are unlike any other living species. We are entitled to the Earth, and to all its fruits. We *own* the Earth, and we must seek dominion over "every living thing that moveth on the earth."

There is no doubt in my mind that the Israelites (and presumably their agriculturally based neighbors) saw themselves as living outside—or above—the natural world surrounding them. This is an extremely accurate picture of who they in fact were, and what their actual relation to the natural world was. (I of course disagree with their account of how they got there in the first place, but we are discussing not competing versions of history, but rather functional views of the here and now, as seen in both scientific accounts and in religious traditions and concepts of God.)

I am also struck by the courage it took to make that leap—though not perhaps so much by the literal change in ecology—which was, on the whole, a logical improvement in efficiency (though some anthropologists and historians have noted how prone to famine agricultural societies through the intervening millennia have been). But there must have been a strong emotional aspect to seeing oneself as outside of nature—the local universe—that must have been deeply disturbing, perhaps even downright terrifying.

So, I see the Judeo-Christian God as an abstraction of the older gods of the Edenish ecosystems from which these early agriculturalists had so recently declared their independence. There is nothing inherently new or radical about this suggestion: theologians have long spoken of concepts of God as changing (even evolving!) and are certainly comfortable with the notion that concepts of God are essentially ideas—ideas that can and do change over time, and from culture to culture. But recognizing that concepts of God are ideas in no way jeopardizes the validity of any of those ideas: just because we can point to a correlation between a change from ecosystem spirits to a monotheistic, more abstract God correlated with actual radical changes in human ecology does not invalidate any specific notion of God.[12]

But the correlation does show, once again, deep resonance between the hyperanalytic vision of Western science (i.e., my conclusion that human beings became the first species in the history of life to leave the confines of the local ecosystem when they invented agriculture) and a statement like that in Genesis (i.e., that a Creator-God made humans *very* different from all manner of other creatures).

Scientific Analysis III. Species are, in essence, packages of genetic information—groups of organisms capable of breeding with each other but not, as the overwhelming rule, with members of other species. Entire species are not parts of ecosystems, meaning that species as a whole do not play direct economic roles in nature. *Parts* of species—local populations—do play direct economic roles in local ecosystems, but species as a whole are information repositories—nothing more. Except, once again, our own species: *Homo sapiens.*

No longer playing direct roles in local ecosystems, it turns out that, particularly with our heightened technological skills in communication, all 6 billion of us are interlocked in an extraordinary global economic network that sees the exchange of no less than a *trillion* dollars worth of goods and services every day. There are, of course, other species that have spread around the entire globe (largely through the hand of humans, like some species of rats, or fruit flies), but these species do not form an integrated economic system (the rats in Norway have no connections with those in New York or Hong Kong, except sporadically through their genes; what other rats are eating in those far-flung regions is sublimely irrelevant to rats of the same species living in London or Tokyo).

If *Homo sapiens* has emerged as the first species in the history of life on Earth to be an economic, in addition to a genetic, entity,[13] we must ask, In what larger-scale economic system is *Homo sapiens* operating? The answer is fairly clear: the sum of all the world's ecosystems is the global ecosystem (sometimes called the biosphere, or Gaia). *Homo sapiens* is exerting a direct—largely negative at the moment—economic impact on the world's ecosystems and is the culprit behind the existing—and growing—dilemma of ecosystem degradation and species loss. Only by muting our impact on the world's ecosystems—taking only what we need (like the Ba Mbuti), engaging in sustainable development and conservation practices, and becoming more efficient (including just plain taking less)—can we hope to stem the tide of the extinction event now gripping the Earth's ecosystems and species.

Concepts of God III. The emergence of the economic impact of humanity on the biosphere as a whole is so new that it is perhaps to be expected that no religious traditions independently mirroring the relatively recent scientific understanding of the problem have yet emerged. Yet, in my travels to colleges and universities across America—especially when speaking either on creationism or, more positively and interestingly, on the resonance I see between science and religion on profound ecological issues—I have learned of a growing movement in *conservative* Christian circles, a movement that can only be described as "green."

Key to this movement is—once again—the interpretation of the "dominion" passage of Genesis 1. Environmentalists for the past few decades have tended to point an accusing finger toward this very passage, saying that, in effect, the injunction to "have dominion" over the beasts of the field is really a license to plunder, rob, and rape the planet—to view everything as put there solely for our own selfish needs—and an invitation to keep taking as long as there is anything left.

But "dominion" easily yields to "stewardship"—husbanding (taking care of) all resources, biological and otherwise. This is precisely what I am hearing from many conservative Christian students—students who are not so much reformulating their concept of God, perhaps, as reinterpreting his instructions on how to treat the beasts of the field according to a sense that we are, after all, still a part of Creation—not at all something a cut above or beyond nature.

I have already remarked that not all conservative Christians are politically conservative. It is true that, in the culture wars, the Christian right—and much of the Republican Party (though there are some notable exceptions)—is genuinely hostile to espousing environmental issues, as if improved efficiency and protection of vital resources were literally inimical to big business. But the younger generation of Christian conservatives seems to have no problem whatever melding its faith with a growing sense that, if we do not address our environmental problems, including the degradation of ecosystems and consequent loss of species, perhaps not much of a Creation will remain over which to enjoy dominion.

I realize that this brief account—especially with two of its components looking strictly at the Judeo-Christian tradition—begs the question, Does this resonance also show up when we look at the religious traditions of pastoralists (generally nomadic agriculturalists)? How do the other known ancient religions associated with the early days of agriculture and the rise of cities fit in—the religions of the Egyptians, Assyrians, Phoenicians? How about the later Greeks, or the Romans? And what about the great Eastern religions?

It's about time we look at all these religions as we search for resonance—for a common understanding of the grave threats faced by the world's ecosystems and species. (Indeed, it's my intention to do so in preparation for my forthcoming book, *Who Is This Man They Call God?*) But I think I have said enough to suggest that in this, the second great arena where my chosen profession intersects matters of public interest (creationism is one, the great loss of biodiversity is the other), the shoe is on the other foot: instead of warfare, we find essential agreement.

The tired old creationism debate—mired as it so thoroughly is in the nineteenth century—simply has not prepared us for the kind of positive interaction between science and religion that I see as eminently possible as we enter the new Millennium and grapple with tough environmental issues.

Nor do I think we can afford these stupid culture wars, with people like Phillip Johnson getting upset that his version of God seems threatened because scientists have discovered that life developed over 3.5 billion years ago on the planet and feel that they can explain how that happened through purely natural causes. Nor can we afford the arrogant intolerance of the scientists who claim that their science—evolution in particular—demonstrates unequivocally that there is no God.

In 1916, sociologist James H. Leuba conducted a poll of scientists active around the nation. Leuba found that the percentage of scientists believing in a "God to whom one might pray in expectation of receiving an answer" was lower than that of the general population. He predicted that, over time, increasingly higher percentages of scientists would declare their disbelief in such a God. A few years ago, as the twentieth century began to draw to a close, Larry Witham and Edward Larson repeated Leuba's poll, contacting a thousand people listed in *American Men and Women in Science* and asking precisely the same questions that Leuba had asked. They found that the *58 percent who had expressed disbelief in such a personal God in 1914 had risen to only 60 percent by 1996.* In other words, there was virtually no change: 40 percent of the scientists polled expressed belief in such a personal God.

I was not surprised by their results. I myself may not affect such belief, but I know a number of colleagues who do. The number of religious scientists grows, of course, when one expands belief to a less proactive conception of a Christian God, or, of course, acknowledges that other religions—Judaism, Islam, Buddhism, Hinduism, and so on—are, well, actual religions, equally valid per se as the narrow-minded version of Christianity espoused by many, if not all, creationists. The intolerance for other people's views—for the genuine religious beliefs found among scientists, for the belief in theistic evolution (Johnson's "theistic naturalism," in which God is seen as having created heaven, Earth, and all of life but did so through natural law), for other religions in general—reduces this parlous little culture war down to a fight to have a purely right-wing Christian nation, where everyone speaks English, is free to tote a gun, and maybe preferably is Caucasian. This is a stupid, hurtful little political battle—this creationism stuff—having lost its last vestiges of intellectual content not long after 1859.

It's not Phillip Johnson's personal God that I'm after; it's his political agenda—specifically his desire to see science watered down in the classroom. I've said enough about concepts of God to make it clear that I think all concepts of God are valid: they are, to me, statements—largely accurate and always interesting—of how people see themselves vis-à-vis the universe in which they live, whether that universe is a local patch of forest, an early agricultural nation-state, or the enormous globe on which we live—or even the universe of modern Big Bang theory.

My wife Michelle thinks that God is gravity. I can buy that; after all, as she points out, gravity holds the universe together, and no one understands it completely either. Closer to home, if someone tells me that God is Mother Earth, that makes sense to me as well. *All* concepts of God—including the intensely personal Creator-God some people see in the Bible—have sprung from a sense of the nature of the universe and how we fit into that universe. It is high time to drop the harmful nineteenth-century stuff underlying the culture war and start the business of dealing with the very real world as we find it at Ground Zero, year 2000.

# Creationism as Theater

As a flashy piece of showmanship, the creationist-evolution trial held in Little Rock a few months back did not match the famous Scopes case of 1925. The Scopes trial, after all, was deliberately staged, contrived by the local citizenry to put Dayton on the map. H. L. Mencken enjoyed a field day and found his various prejudices confirmed, William Jennings Bryan and Clarence Darrow played to the crowds, and even Judge Ralston courted reporters. And there was plenty of "color"—revival meetings, hucksters, and hawkers imparted a carnival atmosphere throughout the proceedings in Tennessee.

Little Rock, in contrast, offered little along these lines—though everyone looked for it. A man showed up in a gorilla suit, people walked around wearing banana buttons, and a crèche lay in the hotel lobby across the street. Most of the relevant symbolism, however, was overlooked. No photographer snapped the "Mon Ark" boat exhibited at the airport, probably because it was so much the obvious product of 20th century technology that its name didn't ring any bells. In 1925, it would have reminded everyone of Noah. Times have changed, and today the young man from the Moral Majority does not spout fire and brimstone but wears a three piece suit and talks quietly, if determinedly, into network microphones.

But on a more significant level, the recent trial was far better drama than Dayton could muster. Dayton gave us fire but little substance. If Little Rock was short on color, it was long on content. For one thing, it was the first such trial in which scientists played a leading role. As for story, it revolved around a carefully orchestrated, if slightly distorted, microcosm of the creationist scene today. Besides scientists, the cast of characters included creationists, prominent members of the Arkansas clergy, theologians, educators, high school science teachers, and of course, lawyers. The plot was simple: a suit brought by the American Civil Liberties Union challenging Arkansas Act 590, which tries to insert biblical literalism into the classroom by calling creationism science.

In the theater of the absurd, plays frequently present an aura of the surreal. Forcing the creationism evolution wrangle into the formalized legal mode of the courtroom evoked just such a feeling of slightly other than real. But creationists are altogether too real and their effects too manifest in the schoolrooms across America to be dismissed as mere actors in a farce. The drama in Arkansas was staged by lawyers, of course, but the effect was nonetheless illuminating: All the words were there, and science—both legitimate and ersatz—received a pretty thorough airing. In a sense, creationists, scientists, educators, and religious personages were like marionettes on strings. Naturally it was the lawyers who pulled the strings, and the sequence of testimony was all carefully geared to this or that legal point bearing on the law's constitutionality: its establishment of religion in the classroom, its abridgement of academic freedom, its vagueness.

But no one watching the events unfold daily had the slightest doubt about what was *really* going on—and here is the greatest source of the nagging feeling that we were witnessing a highly organized and well staged charade. I doubt that anyone in the courtroom really believed the Arkansas attorney general's contention that "creation-science" as defined in the act, whatever its resemblance to the *Genesis* narrative, is not religion but purely science. Certainly the fundamentalists in attendance saw the issue clearly: In their eyes, Act 590 restored God to the classroom, if not supplanting at least receiving equal time with that "atheistic" doctrine of evolution. This, of course, was what the ACLU was trying to show—but the state's defense of the act deviated far from everyone's common sense understanding of creationism and its inspiration in fundamentalist Protestantism and biblical literalism. The attorney general went so far as to produce a theologian to testify that the notion of God is not inherently religious (and this from a man who freely admitted his belief that UFOs are "fallen angels"), divorcing the proceedings from reality to impart, once again, a feeling of theater.

It is easy to understand why the defense especially feared the scientific testimony: The attorney general and his staff were simply outmanned, outgunned by the scientists' attorneys. After all, the educational backgrounds of the attorney general's lawyers at least conceivably made them conversant with many of the basic terms of history and religion—even sociology. But science? Evidently not. Like lambs to the slaughter, one by one the cross-examiners arose from their seats to try to do battle with Michael Ruse (a philosopher and historian of science at the University of Guelph, Ontario), Francisco Ayala (a geneticist at the University of California, Davis), Brent Dalrymple (chief of the western section of the United States Geological Survey and an expert on isotopic dating), Stephen Jay Gould (a Harvard paleontologist and evolutionary theorist), and Harold Morowitz (a biophysicist at Yale University). By the time we got to Dr. Morowitz, the scene had been reduced to low comedy: The fourth member of the attorney general's staff, blatantly ignorant yet doggedly determined to shake, if not break, the placid professor, was hammering away on the second law of thermodynamics. Morowitz, who had already testified in great detail on the thermodynamics of biological systems, calmly and repeatedly replied that his expertise lies in biophysics and that he could not comment on the work of astrophysicists. Undeterred, the lawyer pressed Morowitz to admit that he surely must have some opinion on "astrophysicism." Such was the strength of the attorney general's counterattack on the ACLU's science case. Such is the general strength of "creation science." Moments like this provided the dramatic high points of the trial—isolated peaks of comic buffoonery.

The basic content of science is not in itself the stuff of drama. But what scientists do in the course of their work often is—witness the shenanigans in the competitive rat race for Nobel Prizes, or the more serious issues raised when the scientific endeavor takes on some larger social significance. Creationists have spuriously convinced many citizens that huge hunks of science are antithetical to their religious beliefs—so once again "evolution scientists" have been forced into the limelight, to argue that theirs is the stuff of true science, and that "scientific creationism" is pseudoscience.

Though there will be more acts in the courtroom variety of creationist drama, Judge Overton's decision in Little Rock against statute 590 has taken a lot of wind out of creationism's legislative sails. There will be fewer eager faces in the audience when the next curtain rises—in New Orleans later this year where virtually the same bill will once again be tested in a federal court. But the more prosaic, yet more pervasive, local mini theater version of the same show enacted at school board meetings shows no signs of flagging interest.

Good drama explores and mirrors the fabric of human existence. The creationist ploy—that biblical literalism can be sold as science—is merely play-acting of the nursery school sort. It tells us little about ourselves beyond the depressing point that scientific illiteracy is pandemic—something many might have suspected, but few saw as the enormous problem it really is. Creationism fails as sustaining theater on all counts: It is bad farce and bad soap opera. And in itself it lacks the essence of true tragedy. But the real and potential effects this nonsense has on our school kids are not at all good. And that is the real tragedy.

Niles Eldredge
*Science,* April 1982, pp. 100-101

# The National Center for Science Education

The National Center for Science Education, Inc., (NCSE) is a not for profit, membership organization that defends the teaching of evolution in the public schools. Most of its members are scientists, but many are citizens with an interest in science and education; many other members are concerned with the church and state separation issues engendered by the efforts to "balance" the teaching of evolution with the presentation of religiously based views.

And indeed, in many states and local school districts, there are ongoing efforts to eliminate or discourage the teaching of evolution, or to present religious views as science. The calls for information and assistance to NCSE steadily increase. There are now two creation science organizations funded at about $5 million per year, another major creationist organization funded at over $1 million, at least a half dozen minor ministries focusing on grassroots evangelism against evolution—and periodic antievolution assaults from television and radio evangelists listened to by millions.

Grassroots efforts require grassroots responses, and the National Center for Science Education is the only national pro-evolution organization with this grassroots focus. The NCSE recognizes that the testimony of a local science teacher or college professor at a school board meeting is more influential than a statement from a nationally-prominent scientist—and the staff of the NCSE can help that local person be effective. NCSE prefers to work behind the scenes, letting our members be the "ground troops".

NCSE provides analyses of creationist arguments, background on legal and religious aspects of the controversy, and information on science education. Unfortunately, antievolutionism will not be solved by simply throwing science at it. A clear understanding of scientific aspects of the controversy is essential, of course: if creation science is scientific, then it deserves a place in the curriculum. It is not scientific; it does not belong in the curriculum. But showing that creation science is bad science will not ensure that evolution will be taught: this requires assuaging people's fears that acceptance of evolution requires the abandonment of faith. So NCSE also must provide information beyond the scientific: connections with other interested citizens and groups—including religious denominations that accept evolution, and which do not want sectarian religion taught in the public schools—and advice on how to write effective letters to the editor and op-ed pieces, and on other media relations.

Calls for information to support the teaching of evolution and/or to counter antievolutionary approaches come from many directions, including state boards of education, state departments of education, committees charged with selecting textbooks, local boards of education, individual teachers, and citizens—a wide range of people and institutions. The NCSE has a speakers' bureau of knowledgeable scientists around the country who can help the public understand why evolution is an important scientific subject and why students should learn it. NCSE also maintains a web page with information on evolution, the creation and evolution controversy, and evolution education.

The National Center for Science Education can be reached at:

The National Center for Science Education
P.O. Box 9477
Berkeley, California 94709-0477,

1-800-290-6006
www.natcenscied.org

## 25 Ways to Support Evolution Education

Parents, teachers, and even scientists often ask, "What can I do to help?" Here are 25 practical, effective suggestions. Using this list is likely to inspire you, and as you work out new ideas, be sure to share them with the National Center for Science Education.

1. Donate books and videos about evolution to school and public libraries. (NCSE can help you choose appropriate materials.)

2. Encourage and support evolution education at museums, parks, and natural history centers (by positive remarks on comment forms, contributions to special exhibits, etc.).

3.Thank radio and television stations for including programming about evolution and other science topics.

4. Make sure friends, colleagues and neighbors know you support evolution education and can connect them with resources for promoting good science education.

5. Monitor local news media for news of anti-evolution efforts in your state or community, and inform NCSE—for example, by mailing newspaper clippings.

6. When there is controversy in your community, add your voice: Hold press conferences with colleagues, record public opinion announcements, and send letters or editorials supporting evolution education to local newspapers.

7. Ask organizations in your community to include questions about science education in questionnaires for school board candidates and other educational policy makers.

8. Share your views with school board members, legislators, text-book commissioners, and other educational policy makers.

9. Share NCSE publications with concerned citizens, educators, and colleagues.

10. Link your personal or organizational web-site to www.natcenscied.org

11. When you see a web site that would benefit by linking to NCSE (for example, a science education site), write to their webmaster suggesting the new link to NCSE (www.natcenscied.org).

12. Encourage professional and community organizations (like the PTA) to give public support to evolution education. Send copies of their public statements to NCSE.

13. Join the NCSE.

14. Give gift subscriptions to Reports of NCSE to friends, col-leagues, and libraries.

15. Take advantage of NCSE member benefits like discounts on book purchases and car rentals.

16. PARENTS: Make sure your child's science teacher knows s/he has your support for teaching about evolution, the age of Earth, and related concepts.

17. PARENTS: Help your child's teacher arrange field trips to nat-ural history centers and museums with appropriate exhibits.

18. PARENTS: Discuss class activities and homework with your children—this is often the way communities learn that "creation science" is being taught; or, you may learn your child's teacher is doing a commendable job of teaching evolution.

19. PROFESSIONALS: Inform your colleagues about the evolution/creation controversy and the need for their involvement: for example, by making presentations at professional society meetings, writing articles for organizational newsletters, making announcements on email listserves.

20. COLLEGE TEACHERS: Make sure that your institution has several courses that present evolution to both majors and non-majors.

21. COLLEGE TEACHERS: Create opportunities to learn about evolution outside the classroom: for example, public lectures, museum exhibits.

22. K-12 TEACHERS: Work with your colleagues to create a supportive atmosphere in your school and community.

23. K-12 TEACHERS: Work with colleagues to develop or publicize workshops and in-service units about evolution; take advantage of them yourself.

24. INFORMAL EDUCATORS: Include evolution in signage, interpretation of exhibits, docent education, and public presentations.

25. SCIENTISTS: Share your knowledge with K-12 teachers and students by visiting classrooms or speaking at teacher-information workshops (NCSE can provide tips).

Eugenie Scott
Director
The National Center for Science Education

# Seven Significant Court Decisions on the Issue of Evolution versus Creationism

1. *Epperson v. Arkansas:*
   In 1968, the U.S. Supreme Court invalidated an Arkansas statute that prohibited the teaching of evolution. The Court held the statute unconstitutional on grounds that the First Amendment to the U.S. Constitution does not permit a state to require teaching and learning to be tailored to the principles or prohibitions of any particular religious sect or doctrine. *(Epperson v. Arkansas,* 393 U.S. 97; 37 U.S. Law Week 4017; 89 S. Ct. 266; 21 L. Ed 228 [1968]).

2. *Segraves v. State of California:*
   In 1981, a California Superior Court found that the California State Board of Education's Science Framework, as written and as qualified by its antidogmatism policy, gave sufficient accommodation to the views of Segraves, contrary to his contention that class discussion of evolution prohibited his and his children's free exercise of religion. The antidogmatism policy provided that class discussions of origins should emphasize that scientific explanations focus on how, not ultimate cause, and that any speculative statements concerning origins, both in texts and in classes, should be presented conditionally, not dogmatically. The court's ruling also directed the board of education to widely disseminate the policy, which in 1989 was expanded to cover all areas of science, not just those concerning issues of origins. *(Segraves v. California,* Sacramento Superior Court No. 278978 [1981]).

3. *McLean v. Arkansas Board of Education:*

In 1982, a federal court held that a "balanced treatment" statute violated the establishment clause of the U.S. Constitution. The Arkansas statute required public schools to give balanced treatment to creation science and evolution science. In a decision that gave a detailed definition of the term "science," the court declared that creation science is not in fact a science. The court also found that the statute did not have a secular purpose, noting that the statute used language peculiar to creationist literature in emphasizing origins of life as an aspect of the theory of evolution. Although the subject of life's origins is within the province of biology, the scientific community does not consider the subject as part of evolutionary theory, which assumes the existence of life and is directed to an explanation of how life evolved after it originated. The theory of evolution does not presuppose either the absence or the presence of a creator. *(McLean v. Arkansas Board of Education,* 529 F. Supp. 1255; 50 U.S. Law Week 2412 [1982]).

4. *Edwards v. Aguillard:*

In 1987, the U.S. Supreme Court held unconstitutional Louisiana's Creationism Act. This statute prohibited the teaching of evolution in public schools, except when it was accompanied by instruction in creation science. The Court found that, by advancing the religious belief that a supernatural being created humankind, which is embraced by the term "creation science," the act impermissibly endorses religion. In addition, the Court found that the provision of a comprehensive science education is undermined when it is forbidden to teach evolution except when creation science is also taught. *(Edwards v. Aguillard,* 482 U.S. 578 [1987]).

5. *Webster v. New Lenox School District:*

In 1990, the Seventh Circuit Court of Appeals found that a school district may prohibit a teacher from teaching creation science, in fulfilling its responsibility to ensure that the First Amendment's establishment clause is not violated and religious beliefs are not injected into the public school curriculum. The court upheld a district court finding that the school district had not violated

Webster's free speech rights when it prohibited him from teaching creation science, since it is a form of religious advocacy. (*Webster v. New Lenox School District No. 122*, 917 F. 2d 1004 [1991]).

6. *Peloza v. Capistrano School District:*
In 1994, the Ninth Circuit Court of Appeals upheld a district court finding that a teacher's First Amendment right to free exercise of religion is not violated by a school district's requirement that evolution be taught in biology classes. Rejecting plaintiff Peloza's definition of a "religion" of "evolutionism," the court found that the district had simply and appropriately required a science teacher to teach a scientific theory in biology class. (John E. Peloza v. Capistrano Unified School District, 917 F. 2d 1004 [1994]).

7. *Freiler v. Tangipahoa Parish Board of Education:*
In 1997, the U.S. District Court for the Eastern District of Louisiana rejected a policy requiring teachers to read aloud a disclaimer whenever they taught about evolution, ostensibly to promote "critical thinking." The court wrote, "In mandating this disclaimer, the School Board is endorsing religion by disclaiming the teaching of evolution in such a manner as to convey the message that evolution is a religious viewpoint that runs counter to . . . other religious views." The decision is also noteworthy for recognizing that curriculum proposals for "intelligent design" are equivalent to proposals for teaching creation science. (*Freiler v. Tangipahoa Parish Board of Education*, No. 94-3577 [E.D. La. Aug. 8, 1997]). On August 13, 1999, the Fifth Circuit Court of Appeals affirmed the ruling.

National Center for Science Education

# Notes

## Chapter 1: In the Beginning

I. page 10
Wertheim is the author of two books and many important articles on the relation between science and religion, as well as the writer and presenter of a major PBS documentary on the subject. I have had the pleasure of sharing the podium with her on many college campuses, where we have each offered our different—yet compatible—takes on the relation between science and religion. Her thinking has deeply influenced my own and is reflected in the pages of this book.

2. page 10
I once lectured at a small, conservative Christian college in western Pennsylvania, where I was told by the dean who had invited me that some of the students at the all-campus talk I had given the previous evening probably had been looking for horns on my head—if an exaggeration, perhaps not too fanciful. I asked the dean why he had invited me, and he said to shake the kids up a bit, to let them know there were more views that could be intelligibly expressed than many of them had encountered. On that trip, I also encountered for the first time glimmers of a deep interest in environmental matters that some of the students had begun to develop—my first glimpse of the growing ecological movement within parts of the conservative Christian community.

3. page 10
In March 1984 I was invited to deliver a lecture as part of a series celebrating the hundredth anniversary of the founding of the Jewish General Hospital in Montreal, Canada. At the time, I had been busy battling creationists, and I always considered the likelihood of confronting creationist hecklers in any prospective audience (not that I ever turned down a speaking request for fear

of such hecklers, for forewarned is forearmed). I figured that a Jewish hospital in a basically Catholic city would be probably the last place on the continent I would run into creationists. Wrong! At the end of my talk, I was politely quizzed on what I thought of the correspondence between geological eras and the days of creation as outlined in the first book of the Pentateuch (i.e., Genesis). My interrogator was no Christian fundamentalist, but rather an Orthodox Jew, a doctor on the hospital staff who, for religious reasons, had trouble with the ideas of evolution and deep geological time.

4. page 20

I grew to respect—and even like—Luther Sunderland, who is since deceased. He was a worthy opponent, whom I often debated on TV shows. He was a genuinely talented engineer who built and flew his own airplanes and who was clearly no enemy of physical science or technology. On one program, during a commercial, he turned to the famous Isaac Asimov, another guest on the show, and said something to the effect of, "I don't get it; here you are probably the foremost spokesman for science in America, yet you won't fly on airplanes! Don't you believe in the laws of aerodynamics?" Creationists are right when they (some of them, at least) claim they are not antiscience per se; to which, of course, the only reply can be that to be in favor of certain fields of science (most of physics, say) but not others (evolutionary biology or historical geology) implies a set of criteria that are not themselves scientific.

## Chapter 2: Telling the Difference

1. page 21

Several years after first reading reports of Reagan's remarks, I heard from Luther Sunderland (already encountered in Chapter 1) that he had managed to get through to several of Reagan's speechwriters and that he was proud to take credit for inserting these antievolution remarks into Reagan's campaign rhetoric. But I wasn't prepared to hear what he had to say next: that the scientists he had in mind—the scientists to whom Reagan was alluding in his none-too-mellifluous antievolution comments—were none other than me and Stephen Jay Gould. Sunderland was saying that our idea of punctuated equilibria (see Chapter 4) was antievolution because it seemed to be anti-Darwin.

2. page 21

Creationists, I am saddened to say, are not the only set of people who are guilty of disparagingly pronouncing evolution as "only a theory." No less a source than the editorial page of the *New York Times* (June 23, 1987, p. A30), in an editorial marking yet another court decision on creationism, was guilty of the very same sloppy usage of the term "theory." The *Times* congratulated the U.S. Supreme Court in deciding that a Louisiana law mandating the teaching of evolution in public schools must be "balanced" by the teaching of creation science is unconstitutional. Commenting on the 7 to 2 vote, the *Times* editorial duly noted that only Justices William Rehnquist and Antonin Scalia sided with the state of Louisiana in its argument in defense of the law. But the editorial also stated, "It's true that evolution is only a theory and that some scientists, contrary to scientific method, treat it as dogma." On June 29 of that year, the Times (p. A16) published my anguished rejoinder in their Letters section:

Quantum mechanics. Special relativity. Plate tectonics. All are theories, yet no one, in my experience, ever says they are "only" theories, and that "some scientists, contrary to scientific method, occasionally treat [them] as dogma." Yet that is how you chose to characterize the scientific status of evolution in your June 23 editorial commending the recent Supreme Court decision voiding the Louisiana "equal time" statute.

Evolution is the only non-supernatural explanation of the mosaic of similarity that interconnects the vast array of otherwise so dissimilar forms of life still current in Western thought. If all organisms are descended from a single common ancestor in the remote geological past, tell-tale vestiges of their common origin should show up as traits still shared—in the macromolecules of heredity (DNA and RNA) that unite everything from bacteria to mammals.

That there is one grand pattern of similarity linking all life forms is the overwhelming conclusion of 300 years of research in systematics and paleontology. It is also the practical basis of all non-human biomedical research. Evolution is as well established as any other complex theoretical position in science.

Yes, evolution is a theory—in the sense that it is an idea that explains a tremendous amount of information about the living world. But to say "only" a theory demeans and misrepresents the very nature of the scientific enterprise. And it misses the power of evolution, which has been one of the most fruitful ideas ever introduced in Western civilization.

I explain the scientific details of this letter more fully in this chapter and examine courtroom decisions on creationism in Chapter 7 (see also appendix 3). For more on the grand pattern of relationships interconnecting all life, see Chapter 3 of my *Life in the Balance* (1998)—and if possible visit the "Wall of Life" (my informal name) in the Hall of Biodiversity (opened May 1998) at the

American Museum of Natural History in New York City. The Wall is the only place where the entire diversity of life, from bacteria through fungi, plants, and animals, can be seen in one 100-foot-long sweep in all its beauty, intricacy, and evolutionary relationships.

3. page 23
Of course, we assume the Earth has been round since not long after its inception, probably through the coalescence of smaller planetesimals and even finer bits of galactic debris. This kind of thinking is usually called induction, though when specifically applied in historical geology, it is traditionally referred to as uniformitarianism. I have discussed aspects of the scientific process in greater detail in my recent *The Pattern of Evolution* (1999), in particular examining the meanings and origins of uniformitarianism in geology and paleontology.

## Chapter 3: The Fossil Record

1. page 36
Viruses are essentially strands of genetic material with a protein coat, and as such they are indeed simpler than bacteria. Yet all viruses require the presence of another organism's genetic material to reproduce, and they lack the metabolic pathways to allow them to "live" in anything other than a vegetative state when they are not reproducing. Thus viruses are not independent, free-living organisms. I agree with some biologists who see viruses as escaped snippets of the genetic material of advanced, true organisms—snippets that must reinvade a host's genetic machinery in order to reproduce. Indeed, it seems quite likely that some genes—such as the infamous cancer-causing oncogenes—are essentially viruses that have reinvaded the genomes of host species and have secondarily become permanent residents. Under this line of thinking, viruses that attack humans and other primates are more closely related to their human and other primate hosts than they are to other viruses, such as the famous tobacco mosaic virus of tobacco plants.

2. page 37
I choose the phrase "life itself" advisedly here, for none other than Nobel laureate Francis Crick (one of the codiscoverers of the structure of DNA), in a book by that title (1981), has argued that the origin of life is biochemically sufficiently difficult, hence improbable, that life may well have arisen elsewhere and spread to Earth (perhaps sent by an advanced civilization). Critics, of course, have noted that this scenario merely puts back the ultimate problem of the origin of life to another time and place. I vastly prefer the simpler idea that life is intrinsic to Earth and arose here—at least until such time as we can show otherwise.

3. page 43
See Mark McMenamin's *The Garden of Ediacara* (1998) for a good overview of the diversity of life in the Ediacaran fauna, as well as a spirited account—and defense—of his interpretation of what these fossils represent in evolutionary history.

4. page 46
See my *Life in the Balance* (1998) for a more complete rundown on the evolutionary spectrum of life, including forms of invertebrate life.

5. page 47
Just as this book was in its final stages of preparation, paleontologists announced the presence of primitive jawless fishes in Cambrian rocks from China.

6. page 47
The Burgess Shale has become familiar to modern readers through several important books, most notably Stephen Jay Gould's *Wonderful Life* (1988) and, most recently, *The Crucible of Creation* (1998) by British paleontologist Simon Conway Morris.

7. page 48
Technically speaking, a biomere is a biostratigraphic unit, meaning a sequence of sedimentary rocks recognized by its fossil content. Biomeres represent considerable spans of geological time; that is, they are also chronostratigraphic units.

8. page 48
See, for example, Brett and Baird, 1995. I have had the pleasure of collaborating on some aspects of this work. In addition, it was Carlton Brett who first told me (on a fast-food lunch break while collecting fossils in Middle Devonian rocks in upstate New York) that it was not just my trilobites that showed tremendous stability, interrupted by infrequent, but rapid, bursts of evolutionary change. The entire rich Hamilton fauna, consisting of over 300 named species of hard-shelled invertebrate organisms, showed the very same pattern. Most of the species were there at the beginning of the 6- to 7-million-year-long interval of Hamilton time, and most were still there, in recognizably the same unchanged form, at the end, when most of them abruptly became extinct, many to be replaced by similar descendant species in the next interval.

9. page 55
Paleontologist James Hopson of the Field Museum of Natural History in Chicago has been especially eloquent in expressing the fruits of his research on the evolution of mammals from mammal-like reptiles, providing one of the best antidotes to the tired old creationist claim that the fossil record reveals no transitions between "major kinds."

10. page 56
See Ian Tattersall's lucid account of human evolution in his *Becoming Human* (1998) for an entrée into the entire field of paleoanthropology.

11. page 57
The morning after I appeared in a debate with creationist Phillip Johnson at Calvin College in Grand Rapids, Michigan, I attended an informal meeting of science faculty. Though I was the evolutionist, the science faculty at this conservative Christian college (Calvin College is affiliated with the Christian Reformed Church in North America) are extremely professional: not only was I treated cordially, but many faculty members made it plain that, whatever their personal feelings about evolution may be, they know that evolution is a bona fide scientific concept—and they have great respect for science, as they themselves are professional scientists. Rather, it was Johnson with whom they had a bone or two to pick, since Johnson apparently cannot understand why, as a leading Christian conservative intellectual, he is the darling of much of fundamentalist and evangelical circles *except* science faculty at some conservative Christian schools.

In any case, as we were waiting for Johnson to show up, we talked about the program the preceding night. I'll never forget one faculty member sitting next to me (I believe he was a physicist), who said—in reference to the series of slides of fossil human skulls I had shown—"Boy, you really went for the jugular!" That's really it: if we evolutionary biologists would only stop brandishing the fossil record of human evolution as one of the very best examples of evolutionary change through time, the creationists would be deliriously happy, and all but a few diehards wouldn't give a damn what we said about trilobites, dinosaurs, or horses!

12. page 57
The ancient ecosystems of eastern Africa are well preserved in sediments that have accumulated in vast down-dropped basins. The exact same story is now emerging in South Africa—this time from a series of caves.

13. page 58
The so-called Taung child, the first discovered specimen of *Australopithecus africanus*, and thus the type specimen of the species, may have been as young as 2.3 million years old. South African lime deposits are difficult to date with precision (no suitable radiometric techniques are available), and, in any case, the limeworks at Taung have long since been quarried out completely.

14. page 59
Paleoanthropologist Ian Tattersall, my colleague at the American Museum of Natural History, makes the case that more than one early species of *Homo* was present in eastern Africa between 2.5 and 2 million years ago. For simplicity's sake, I mention here only the most familiar of these: *Homo habilis* (literally, "handy man"), first discovered and named by Louis Leakey. The most famous specimen of this species, dated at about 1.9 million years, was discovered by Leakey's son Richard. (See also Tattersall and Schwartz, 2000).

15. page 60
The dates given here for the probable time of evolution of *Homo sapiens* are derived from analysis of mitochondrial DNA diversity patterns in modern humans (the so-called Eve hypothesis). See Chapter 1 in my *Life in the Balance* (1998) for an ecological evolutionary account of human prehistory, detailing the crucial events in the physical environment that underlay so much of human evolution. How humans eventually came to declare their "independence" from local ecosystems (via the invention of agriculture beginning some ten thousand years ago) is especially germane to understanding the origins of settled existence (i.e., villages, towns, and cities) and complex social structures. It is also central to understanding how concepts of God have changed as the human ecological niche has changed, as I develop in the final chapter of this book and explore much more fully in my forthcoming book, *Who Is This Man They Call God?*

## Chapter 4: What Drives Evolution?

1. page 62
Genetic cloning is simply the latest phase of human manipulation of the gene pools of plant and animal species for agricultural, biomedical, or simply experimental reasons. Artificial selection of genetic properties began with domestication of animals and plants at least ten thousand years ago, with the emergence of agriculture in a number of different areas (though archeological evidence suggests that certain species were domesticated even earlier—thou-

sands of years before the Agricultural Revolution). Genetic cloning is simply a high-tech and more direct way of manipulating the genes of organisms.

2. page 63
I once gave a talk at a meeting of the New York metropolitan chapter of the scientific "fraternity" Sigma Xi, at an IBM research center. Most of my audience consisted of physicists, chemists, mathematicians, and computer experts. I was describing some of the current debates within evolutionary biology, and in so doing I made probably the most naïve statement I have ever uttered in a public forum: I said, in effect, that the fact that evolutionary biologists tend to argue so much with one another probably sounds strange to people working in the physical sciences. That remark was met by an embarrassing gale of laughter. The audience thought it was hilarious that I would suggest that physicists, chemists, and mathematicians routinely agree on their formulas and other expressions of their scientific conclusions. This was years ago, and I was guilty of assuming, as many people routinely do, that the physical sciences are more precise, and therefore less prone to argumentative discourse and profound disagreement. Embarrassing as the episode was, it taught me an extremely valuable lesson: all areas of serious, systematic human inquiry, certainly including all areas and disciplines of science, are constantly and as a matter of course replete with disagreement—over experimental protocols, observations, and of course their interpretations. Science proceeds by matching up our perceptions of natural phenomena with our thoughts about the nature of these phenomena, and though the essential creativity underlying this work resides within the individual's mind, nonetheless the growth of knowledge (in any field) comes through the collective discourse of many minds—minds that often cannot agree on the nature of those phenomena, let alone their interpretations.

3. page 63
Science writer John Horgan's book The End of Science (1996) posits that all the major discoveries about the natural world have already been made. This conclusion, naturally, upsets many scientists, who do not like to think of themselves as merely dotting the i's and crossing the t's—i.e., relegated to a subordinated mop-up role. Of course, there is something to Horgan's thesis: nothing, for example, more fundamental in the realm of evolutionary biology than the very principle that all life has descended from a single common ancestor is, almost by definition, likely to be discovered in the future. Likewise, natural selection—so abundantly confirmed by experimentation, mathematical and other forms of theory, and observation in the wild—as the central molder and shaper of organismic adaptations, is unlikely to be supplanted or matched by another, as yet undiscovered, evolutionary process. Yet it has been my con-

tention throughout my career that we have a long way to go toward under-standing how and when natural selection acts to shape evolutionary change—what I have called the context of evolutionary stasis and change. There is still much to be learned, as this chapter endeavors to point out.

4. page 64
And differ it does. Dawkins sees evolution essentially as an active process: the direct result of competition among genes for representation in succeeding gen-erations. Darwin's version, to my mind, is infinitely preferable: it sees natural selection as a passive recording of what worked better than what in a world of finite resources—where not every organism born within a local population can possibly expect to survive and reproduce. I have explored the significance of these differences in the fundamental meaning of "natural selection" elsewhere (Eldredge, 1995, 1999, and technical articles cited therein).

5. page 66
Philosopher David Hull has pointed out that the notion of natural selection received more criticism than any other aspect of Darwin's evolutionary ideas—up to and including the very idea of evolution itself. Thus it cannot be literally true that the concept of natural selection was the linchpin of Darwin's success. I tend to agree, since I am convinced that it was the mountain of evidence *in the form of patterns* that was primarily responsible for Darwin's immediate and lasting success. Nonetheless, because the notion of natural selection has been so completely verified and upheld through experimentation, field observation, and mathematical theory for so long—and because it remains the core aspect of our understanding, not only of the evolutionary origin and further modifi-cation of adaptations, but also of the great pattern of stasis (i.e., nonchange) itself—Darwin's development of the notion in the opening chapters of his book, added to the power of all the patterns he presented, must have created an ineluctable one-two argumentative punch that did indeed carry the day for establishing the scientific credibility of the very idea of evolution.

6. page 67
Darwin (and Alfred Russell Wallace), upon reading Thomas Malthus's *An Essay on the Principle of Population* (1798), realized that reproduction within all species (Malthus confined his gaze to human beings) is inherently geometric: if, say, every pair of sexually reproducing organisms produced two offspring, population size would grow by leaps and bounds. Darwin used slow-breeding elephants to dramatize the point—relying on a Victorian phrase still very much in use when he pointed out that the world is by no means "standing room only" in elephants, which by now it must be *unless some factor(s) were limiting the size*

*of elephant populations.* We now realize that it is the productivity (food and nutrients) available in local ecosystems that, along with predation, disease, and physical factors, limits sizes of local populations and thus, additively, of entire species. That the human population has exploded from an (estimated) 6 million to over 6 billion in the brief ten thousand years since the Agricultural Revolution is graphic confirmation of this very point, for it was the invention of agriculture that removed human beings from their primordial position within local ecosystems, thus removing the Malthusian limit to human population size. As we shall see in Chapter 7, in my view this radical shift in human ecology had enormous implications for the very concepts of the gods or God.

7. page 67
This is the significance of the famous cloning of the sheep Dolly in 1998. the genetic information that produced Dolly came from an udder cell—not an egg or a sperm, or indeed the nucleus of a fertilized egg mechanically removed from another sheep.

8. page 69
Occasional reports published in the scientific literature claim the experimental verification of the inheritance of an acquired character; perhaps the most discussed such instance in recent years involves the acquisition, and subsequent transmission, of immunological tolerance. Though most biologists are unwilling to go so far as to resurrect Lamarck's notions of evolutionary change, in the face of a few apparent exceptions to Weismann's doctrine (and, in any case, not all biologists accept such examples as bona fide), it is nonetheless interesting, and utterly characteristic of the process of science, that one of its most solid underpinnings Weismann's doctrine—is still occasionally called into question.

9. page 69
Actually, the word "mutation" was first introduced to the scientific literature on evolution by the German paleontologist W. Waagen, who used the term in the 1880s to refer to discrete, stable stages that he saw in some Mesozoic ammonoid lineages he was studying.

10. page 70
For example, Henry Fairfield Osborn, paleontologist as well as President of the American Museum of Natural History, coined the term "aristogenesis" for his idea that evolution arose from innate tendencies within species to become ever better. Organisms are superior because of innate genetic propensities, or so Osborn thought. Himself a wealthy man—an American "aristocrat"—Osborn

actively supported right-wing eugenics endeavors, and some of his writing was used in support of Nazi propaganda.

11. page 71

The concept of genetic drift was introduced by Sewall Wright, who showed that, under certain circumstances, alleles (variant versions of a specific gene) could become "fixed" in a population without—or even in spite of—natural selection.

12. page 78

Indeed, the work of paleontologist Phillip Gingerich and others has recently produced convincing evidence from the Indian subcontinent of transitional mammals—intermediates between terrestrial predecessors and true whales. Simpson knew—because evolution predicts—that intermediates must always have existed. His mission, as he saw it, was to explain why it is characteristically so difficult to actually *find* intermediates in the fossil record—a situation his theory of quantum evolution was originally conceived to address.

13. page 82

For example, the English molecular biologist Gabriel Dover propounded his ideas on genetic drive, in which certain DNA sequences could alter the sequence of corresponding genes on other chromosomes; Canadian geneticist W. Ford Doolittle and his colleagues suggested the notion of selfish DNA; and Motoo Kimura and other geneticists developed the notion of neutral, or non-Darwinian, evolution—stressing the fact that many alternate genetic forms are selectively neutral (meaning equally viable). These ideas began to surface in the 1960s (neutral theory) on up through the 1980s. None are any longer seen as controversial—since the intricacies of the molecular biology of the gene and its workings offer badly needed deeper understanding of the processes of heredity—but not of phenomena at the levels of the population and higher, such as natural selection, speciation, extinction of species, and so on.

14. page 82

I say that the advent of molecular genetics prompted the emergence of this line of thought not so much because Dawkins's main emphasis concerns competition among genes for representation in succeeding generations, as because the development of molecular biology posed a challenge to more tradition-minded geneticists. This was apparent to me at a meeting on macroevolution held at Chicago's Field Museum of Natural History—where much of the discussion focused on such issues as punctuated equilibria (which I discuss later in this chapter); nonetheless, most of the attention of the population geneticists present at the meeting focused far more on the few molecular biologists present than on those of us who were paleontologists. To put it another way, the development of molecular biology seemed to pose a threat to population geneticists back then—a threat that has dissipated because graduate students in all walks of evolutionary biology (even paleontologists) are routinely trained in the basic laboratory procedures and theoretical constructs of molecular biology.

15. page 82
Hamilton's work is often said to have been a formal extension of work done by Ronald Fisher—supposedly on an envelope in an English pub—a half century earlier.

16. page 83
See, for example, John Endler's *Natural Selection in the Wild* (1986) and John Thompson's *The Coevolutionary Process* (1994) for excellent discussions and entrée into the vast literature of modern population-level evolutionary biology.

17. page 84
See my book *The Pattern of Evolution* (1999) for a discussion of the causes of stasis and details of the idea of punctuated equilibria, including (in Chapter 1) a discussion of the genesis of the idea based on the perception of four basic patterns in these trilobite data.

18. page 87
See my book *The Pattern of Evolution*, as well as Eldredge, in press-a, for a much more detailed description of this "sloshing bucket" model of the evolutionary process.

## Chapter 5: Creationists Attack: I

1. page 90
Creationism is a social rather than an intellectual (let alone *scientific*) issue, and (as we have already seen and will encounter in greater detail throughout the next several chapters) as such it has of course been prominently displayed in courtroom battles (and movies depicting such battles, such as *Inherit the Wind*). Less well known is the penchant that some lawyers have had over the years of applying their legalistic argumentative skills to debunk Darwinism, acting very much as if the debate between evolution and creationism can be handled in much the way a traditional debate—or a courtroom prosecution—is held. I am not the first to point out that lawyers arguing for one side against another in a

case are notorious for wanting to establish the truth of the matter only insofar as their side can prevail or at least reach the most favorable outcome possible. They aim to persuade a judge and jury of the rightness of their position, and the justice system presumes that, by such a process, the actual truth will emerge. At least, that is the idea.

In my professional lifetime, two such lawyers have taken up the antievolution cudgels, and in so doing they attracted not a little notoriety. The first was Norman Macbeth, who I met several times at the American Museum of Natural History and who authored *Darwin Retried* (1971)—once erroneously, if hilariously, listed in a sales catalogue as *Darwin Retired*. To my knowledge, Macbeth always denied he was a creationist, meaning someone who was opposed to evolution on religious grounds. Rather, he steadfastly maintained that his was purely an intellectual interest, and that Darwinism had many holes that could be exposed through counterscientific evidence and, perhaps especially, the cold hard stare of a dispassionate lawyer trained to get to the heart of the matter. Whatever his motivations may have been, though, Macbeth trotted out many of the same objections to evolutionary biology that creationists always do—and added none of his own.

The more recent and better-known efforts of lawyer Phillip Johnson (encountered briefly already in this book and in some greater detail in Chapter 6) are very much of the same stripe, but with a single great difference: Johnson readily admits that it is his belief in a personal Christian God, and the implications he sees (as others have before him) for a materialistic, godless universe latent in the very idea of evolution, that motivates him to wield his lawyerly legerdemain against evolutionary biology. Here I give him credit, since Johnson reflects what I see as a growing tendency through the 1990s for creationists to admit their religious convictions and hence motivations. Hiding behind a deceitful cloak of scientific creationism, as was done in the 1960s through the 1980s for the purposes of circumventing the establishment clause of the First Amendment of the U.S. Constitution—in other words, creating the pretense that creationism was motivated solely by the quest for pure scientific understanding and its communication to children in the classroom—seems no longer as fashionable as it so recently was. In rediscovering their roots and openly acknowledging their underlying religious convictions, creationists have begun to abandon their hope for ultimate legal imprimatur—the efforts of Justice Antonin Scalia of the U.S. Supreme Court to the contrary notwithstanding (see Chapter 7).

I also note that, later in this chapter, I rely not on a creation scientist but rather on yet another lawyer (Wendell R. Bird), who produced the clearest, most succinct statement of the fundamental tenets of scientific creationism—the statement that bridges the gap between openly acknowledged religiously based creationism and "scientific" creationism on the one hand, and the language defining "creation science" in Arkansas Act 590 on the other.

That being said, I reiterate my opening point: I can find nothing truly new in the antievolution rhetoric of creationists, regardless of whether it is cast in the cloak of scientific creationism or in the more recent and more open terms of a Christian lawyer merely arguing the case against evolution.

2. page 90
Phillip Johnson's attack on philosophical naturalism is an exception, since he believes that, if his version of a personal God—a God watching over everything and especially everyone, a God capable of doing everything from making mountains to creating humanity—really exists, then science, if it is to be true to its basic quest to describe the "furniture" of the universe and especially to understand its cause-and-effect workings, perforce *must* take God into account.

3. page 91
This either-or aspect of creation versus evolution was nicely displayed by an utterly coincidental (I presume) juxtaposition of two television shows aired simultaneously in the early 1980s. I used to like to watch Garner Ted Armstrong, that handsome and articulate electronic preacher, now, as far as I know, no longer in the TV business. Armstrong would occasionally devote one of his shows to evolution. The incident in question arose as I was idly cruising the TV channels and came upon Armstrong's gentle but firm rejection of evolution as a plausible explanation of nature's beauty and complexity. Up on the screen popped a film clip showing thousands of silvery grunions—small Pacific Coast fish—flopping around on the beach in Santa Barbara in the moonlight. The grunions were engaged in their annual reproductive rites. Such unusual and complex behavior!, marveled Armstrong. Only a Creator, an almighty God, could have fashioned a fish with such a remarkably intricate reproductive style.

Having gotten the message, I moved along four channels and found an episode of Jacob Bronowski's *Ascent of Man* series on the PBS network. Lo and behold, Bronowski was in the throes of his segment on evolution, and I was treated to my second view of Santa Barbara grunions within five minutes. But what a difference in the moral! How marvelous the process of evolution must be, mused Bronowski, to produce such an intricate pattern of behavior in this little fish!

The unwitting grunions had become a foil for each of the two dominant, competing explanations of how life has come to be as we see it today. Both were frank appeals to the viewer's credulity; both asked us to ask ourselves, What do I believe? I decided to skip the whole thing and watch a movie. But the coincidental Armstrong-Bronowski "debate" did show that, in the United States at least, there really are only two competing explanations for how the Earth and life came to be as we find them today.

4. page 107

Brent Dalrymple, formerly director of the western branch of the U.S. Geological Survey in Menlo Park, California, testified at the creationism trial challenging Arkansas Act 590 in late 1981. Dalrymple performed many experiments to test whether or not extremes of heat and pressure could alter isotopic decay rates and found, not surprisingly, and as atomic theory predicts, that he could *not* alter decay rates experimentally.

5. page 109

At this writing (summer 1999) some astronomers have just revised downward the estimate of the age of formation of the universe, to some 13 billion years. Science marches on, and we have heard the last word on very few issues. Still, it is reasonably certain that, whatever the final say on the exact age of the universe might be (if indeed there ever is such a "final say"), the order of magnitude will be in the range of something over 10 billion years, and a good bit less than 20 billion years, meaning that the basic dimensions of the ballpark will not change.

6. page 114

Skeptical commentators on creationism have understandably had a field day with creationists' attempts to explain modern distribution patterns of plants and animals as the outcome of dispersal from Noah's Ark on Ararat a few thousand years ago. How to get two of each of 10 million species on the ark? How to feed them? And then there is the ever popular sanitation problem: how to clean out what would have surely dwarfed the Augean stables of classical Greek mythology? And so on. Yet the ark story, so easy to poke fun at, is really no crazier than trying to explain the entire sedimentary rock record on the face of the Earth as the result of a single flooding event.

We should all recall that the "world" to the early agriculturalists of the Middle East—wellspring of the writings and religious traditions that have, in one particular extreme form, led to creationism—was limited to the immediate surroundings. Geologists have found some evidence of physical events that may be correlated with events recounted in some of these ancient writings—including, of course, the Old Testament of the Christian Bible. For example, if the geological evidence showing that the Black Sea was a freshwater lake until 5600 b.c.—when the rising level of the Mediterranean caused it to spill over into the Black Sea—is true, as geologists William Ryan and Walter Pitman (1998) report, this event indeed might be the source of the flood stories recounted in Genesis and other ancient documents. The ancients who recorded such events can be forgiven for describing the events as worldwide, for the Middle East was their entire world. What is intolerable is that modern creationists—after geologists tried and failed to make the same claim in the nineteenth century, when geology was in its infancy—still insist in the year 2000 that the Great Flood produced the entire sedimentary rock record.

## Chapter 6: Creationists Attack: II

1. page 118
Biologists at this writing are still studying peppered moths, and the details of the traditional evolutionary account are under critical scrutiny. For this discussion, because *both creationists and evolutionists have agreed on the details in the past,* that the evolutionary story may well in the end be revised does not matter: at issue here is the difference in interpretation between creationists and evolutionary biologists on agreed-upon "facts."

2. page 124
At this writing (summer 1999), much is being made in the popular press of patterns of "cultural" variation between different local populations of African chimpanzees. One of my colleagues (Stephen Jay Gould, writing in the *New York Times)* elicited several indignant letters when he claimed that this pattern helps strengthen the case for continuity between chimpanzees and humans—the overall differences becoming more of degree than of kind. Though I see his point (and, as he acknowledges, such variations in behavior, including the use of simple tools, have long been recognized in other species, such as macaques, as well), I find the definition of culture here to be rather overly encompassing: repeated use of simple tools—e.g., sticks grabbed for the purpose of prizing termites from their nests—strikes me as rather a far cry from the development of learned manufacture (hence cultural traditions of copying style) of even the simplest of the hominid stone tool traditions. Even more to the point, I agree with one of the letter writers, who noted the one great gap—inherently and in principle, "bridgeable," but not in actuality so between humans and any other form of animal life known: the *consciousness* that we humans have. As the writer of the

letter pointed out, human behavior is purposive in the sense that it is con-sciously pursued for survival. All evidence suggests that behavior in all other extant species is instinctive. In other words, I do not think that even the wisest chimp knows, and can therefore contemplate, the fact that it will die someday.

That being said, I must remark that though we have a wonderful human fos-sil record that shows the continuity in anatomical terms between humans and apes—and one that, as well, preserves in outline form the development of human material culture over time—it is a real pity that, unlike the scenarios underlying Arthur Conan Doyle's *The Lost World* (1912), or the more recent movie *Quest for Fire*, we do not have still on Earth with us surviving elements of all the species of our own ancestry that have lived over the past 4 or 5 mil-lion years. Is the origin of consciousness indeed one of degrees—with elements being added as new species (with larger brains and, ultimately, more compli-cated behaviors, including ever more sophisticated material cultural technolo-gies)? *Or*, did human consciousness arise relatively recently, as my friend and colleague anthropologist Ian Tattersall of the American Museum of Natural History has concluded in *Becoming Human* (1998), on the basis of cave paint-ings and other intricate works that began long after our own species had evolved in Africa? If the latter, then perhaps the patterns of geographic varia-tion in simple tool use and other aspects of chimp "culture" do connect on a sliding scale with human material cultural traditions—with the further impli-cation that consciousness itself is not a prerequisite for the development of cul-ture or perhaps even cultural traditions, as exemplified in the toolmaking traditions of the Paleolithic.

Make no mistake: I believe as Darwin did (as opposed to his codiscoverer of natural selection, Alfred Russell Wallace), that human consciousness evolved (probably as a mechanism for survival in a social setting; inklings about what may be going on in a fellow band member's mind, after all, are not only use-ful but best sought by consulting one's own inner thoughts—as biologist Nicholas Humphrey was among the first to point out). But the reason why we humans have become the very first species to stop living inside local ecosys-tems stems from our consciousness and is, I still firmly believe, unique to us.

3. page 130
I have in mind here the then emerging, extremely valuable theoretical approach to systematics (i.e., the study of relationships among organisms and their classification) originally known as phylogenetic systematics, but much more commonly known as cladistics. Cladistics was first formulated by the German entomologist Willi Hennig, but it was actively developed and expand-

ed by a group of American and British systematists (centered largely at the American and British Museums of Natural History) beginning in the late 1960s. The aim was to sharpen the rules of analysis of relationships and bring them into line with principles of scientific testability.

One early realization was that the statement "these two species share features not seen in any other species" is a *positive* statement of evolutionary relationship that is susceptible to falsification—if, that is, other species are discovered later with the same feature, or if still other characteristics show that the species are more closely related to still other species. With ancestors, the situation is trickier because an ancestor will lack at least one of the (advanced) features of its descendant: the ancestor must in all respects be the same as, or more primitive than, its putative descendant—a harder proposition to falsify.

For the record, just because cladistics decrees that there are methodological problems concerning the positive demonstration of ancestors of course *does not mean that that there were no ancestors.* Though some cladists have been content to leave it at that—in any case having no particular, demonstrated interest in how the evolutionary process works—does not mean that some of us who adopted the principles of cladistics in our systematic work did not continue to work with the concept of ancestry and descent. I am the quintessential example: the very notion of punctuated equilibria is based on my tree that reconstructs patterns of ancestry and descent among a series of phacopid trilobites from Middle Devonian rocks in eastern and central North America. The tree was based on a cladogram that I drew up first, on the basis of cladistic principles. I then added information of geological distribution to come up with the tree, and I have always acknowledged that the cladogram on which the tree is based underlies the construction of the tree, and that the tree is therefore a step further removed from certainty than is its underlying cladogram, which of course itself remains a hypothesis subject to further testing. This, of course, is the way science works. Nothing is certain.

In an early paper written with my colleague Ian Tattersall (Eldredge and Tattersall 1976), we explored the differences among cladograms, trees, and the more elaborate scenarios (e.g., complex hypotheses on, for instance, why hominids left life in the trees to assume upright, bipedal gait for walking across the African savannas).

4. page 130
For the record, Sunderland then says, immediately after this quoted sentence, "Anyone, however, can gain access to the original typed verbatim interview

transcripts which were prepared for the New York State Education Department by going to any public library in the United States and asking for the ERIC Document Reproduction Service microfiche ED 228 056," *Darwin's Enigma: The Fossil Record*. The printed version in my possession, then, does not have the full transcripts.

5. page 130
Indeed, I must thank Wesley R. Elsberry, a student in the Department of Wildlife and Fisheries Science at Texas A&M University, who brought this matter of Sunderland and my purported statements about horse evolution to my attention.

6. page 132
I hope I wasn't literally as incoherent as this quote suggests—all of a sudden changing the discourse from horses to rhinos. But it is the case that, at least dentally, the earliest members of the rhino and horse lineages are considered very difficult to tell apart. Rather than representing variation within created kinds, the divergent horse and rhino lineages are far more different in later stages of their evolution than at the beginning. Both are members of the perissodactyl lineage of mammals.

7. page 132
This is a test: I have deliberately written this sentence with the full expectation that it will inevitably be picked up and put in creationist ravings on the Internet and in what passes for creationist "literature." I am equally certain that the following several sentences, in which I reveal what was going on with my interviews with Luther Sunderland and Sylvia Chase, will *not* appear with it.

8. page 134
I owe the title of this section to evolutionary biologist Richard Dawkins, who once remarked, "When I open a page of Darwin, I immediately sense that I have been ushered into the presence of a great mind. I have the same feeling with RA Fisher and GC Williams [two twentieth-century evolutionary biologists encountered in Chapter 4]. When I read Phillip Johnson, I feel that I have been ushered into the presence of a lawyer" (Dawkins, 1996, p. 539). Dawkins's remarks are the opening lines of a "Reply to Phillip Johnson," just one of many articles written by Johnson, some of his cohorts, and several of his critics, in a special issue of the journal *Biology and Philosophy* that I will have occasion to mention again in my treatment of the Phillip Johnson phenomenon.

9. page 134

For example, in his book *Reason in the Balance* (1995), Johnson writes, "Belief in God may persist, particularly in people who have only a shallow understanding of science, but can never have more than a tenuous standing in the world of the mind. Science can step forward at any time and employ its prestige to take control of any subject, even subjects inaccessible to empirical investigation like the ultimate beginning itself. Metaphysical statements by prominent scientists are accepted in the press and throughout public education as advances in scientific knowledge; contrary statements by theologians or religious leaders are dismissed as 'fundamentalism.' The naturalists hold the cultural power; theists in academic life have to accommodate as best they can" (p. 196).

10. page 135

Johnson does occasionally mention creationists in *Darwin on Trial* (1991). For example, at one point (p. 114). he reproduces a statement by Judge William Overton (presiding judge in the famed Arkansas trial over the Arkansas "equal time" law in 1981), who himself was citing creationist Duane Gish (according to Johnson) as reason to conclude that creationism isn't science because it invokes the supernatural. The judge quotes Gish as saying, "We do not know how God created, what processes He used, for God used processes that are not now operating anywhere in the natural universe. This is why we refer to divine creation as Special Creation. We cannot discover by scientific investigation anything about the creative processes used by God." A page later, Johnson distances himself from Gish in saying, "I am not interested in any claims that are based upon a literal reading of the Bible, nor do I understand the concept of creation as narrowly as Duane Gish does. If an omnipotent Creator exists, He might have created things instantaneously, in a single week or through gradual evolution over billions of years. He might have employed means wholly inaccessible to science, or mechanisms that are at least in part understandable through scientific investigation." Johnson elsewhere makes it plain that he is not a "young-Earth" creationist. Elsewhere in his writings, Johnson makes crystal clear his opposition to the idea that God may have used the "purposeless," "undirected" laws of random mutation and selection to create anything, whether over billions of years or not.

11. page 136

See especially philosopher Robert T. Pennock's "Naturalism, Evidence and Creationism: The Case of Phillip Johnson" (1996b)—plus Johnson's reply and Pennock's rebuttal—all in the aforementioned special issue of *Biology and Philosophy* (see note 8).

12. page 139

Ironically, later while teaching at the University of Louvain, Belgium, Mivart fell out of favor with the church for his pro-evolutionary stance and was eventually excommunicated, in 1900.

13. page 140

Behe often claims that relatively little work—even in such outlets as the *Journal of Molecular Evolution*—is actually devoted to specifying pathways of molecular evolution. Rather, much of the work is devoted to using molecular evidence to assess evolutionary relationships among groups—of mammals, say. There is some truth to the remark, but I will never forget a lecture I attended at the Smithsonian Institution in the early 1970s, at which the biochemist Emmanuel Margoliash invited his small audience to don stereovision glasses to view his slides of the reconstructed models of the intricate molecule hemoglobin—so critical to the transport of oxygen in the blood of animals. The structure of the molecule is different in different animal groups, and the DNA coding sites for the molecule have also changed in the course of animal evolution; both of these issues were ably tackled and discussed in the lecture. Biologists can and do confront the evolution of molecular pathways in their data.

14. page 142

Dawkins, like Provine and some other prominent evolutionary biologists, is proud to proclaim his atheistic proclivities. The very title of Dawkins's book— *The Blind Watchmaker*—an overt reference to natural selection as the biological equivalent of the Creator-God of the Judeo-Christian tradition, is, of course, a deliberate red flag in the face of creationists. The penchant of some scientists—especially those with a wide public following—to agree with creationists like Phillip Johnson that the naturalism of science indeed implies that the Judeo-Christian God does not exist, strikes me as crass and rather stupid. For one thing, it is almost incredibly parochial, saying, in effect, that evolution and a Johnsonesque personal God are the only alternatives in the entire earthly domain, rather than acknowledging that the entire "debate" is a peculiar and bizarrely outdated excrescence of (in its greatest part) English and (increasingly) American dated culture—as should by now be obvious to any fair-minded person who has read this far, and as I'll endeavor to characterize further in the next chapter.

15. page 145

I explore the similarities and differences between biological evolution and the history of human designed systems (specifically cornets) in Eldredge, in press-b.

## Chapter 7: Can We Afford a Culture War?

1. page 150
The concept that anyone *could* have been present "when life was created" is another fine example of inane creationist writing. Note, too, the "theory, not fact" dichotomy—one that, as we saw in Chapter 2, is also meaningless. With trash like this in our textbooks, how can we hope to produce a literate society whose citizens are equipped to deal with the complex science- and technology-related issues of the day?

2. page 153
See Daniel Kevles's excellent book, In the *Name of Eugenics* (1985), for a history of the eugenics movement and the involvement of Francis Galton, Darwin's nephew, in it.

3. page 156
I take this quotation from a feature article of the *National Center for Science Education Reports* (vol. 15, no. 4, Winter 1995, p. 3) entitled "Pat Buchanan Takes on Darwin." Buchanan is a Catholic and thereby an excellent example of the marriage of right-wing political views with antievolution sentiment; the Catholic Church has, in the main, accepted evolution—formally so only very recently, but in a "render unto Caesar" sense, as it has recognized since the late nineteenth century that science and religion are essentially separate domains.

4. page 156
This quotation comes from "Mr. DeLay's Power Play," lead editorial, Sunday, June 20, 1999, p. 14, in the *New York Times, Week in Review.*

5. page 158
At this juncture, it will hardly come as a surprise that no less an important creationist figure than Henry Morris himself, also writing in *Acts and Facts* (July 1980) completed the circle, writing, "Creationism can be studied and taught in any of three basic forms, as follows: (1) Scientific creationism (no reliance on Biblical revelation, *utilizing only scientific data* to support and expound the creation model). (2) Biblical creationism (no reliance on scientific data, using *only the Bible* to expound and defend the creation model). (3) Scientific Biblical creationism (full reliance on *Biblical revelation* but *also* using *scientific data* to support and develop the creation model)" [Morris's italics]. The creation model! Morris is clearly saying that the creationist position can be interchangeably considered religion, science, or a mixture of the two—depending on the intended audience.

On another note, Judge William Overton, in his opinion striking down Arkansas Act 590, noted that Wendell Bird's article in the *Yale Law Journal* in 1978, was a "student note" with "no legal merit." Bird had tried to make the case that the religious aspect of evolution violated students' rights, thereby making it appropriate to teach creation science concurrently.

6. page 158
*Science* has, over the years, on the whole done a rather good job of covering the more public displays of the creationism mess in the United States. In particular, I recommend the news accounts of the Arkansas trial written by a really good and experienced reporter, Roger Lewin. I rely here particularly on Lewin's (1982) account of Judge Overton's decision, "Judge's Ruling Hits Hard at Creationism."

7. page 159
Not surprisingly, lawyer Phillip Johnson (1995) agrees with Justice Scalia of the U.S. Supreme Court that the Lemon test (i.e., the 1971 case opinion providing the three-pronged test of the establishment clause just quoted) should be voided by the Supreme Court because it has led to excessive judicial interference "with attempts by other branches of government to arrive at sensible ways of accommodating religious and secular values in public life" (p. 220).

8. page 160
The estimate of 10 million species on the planet is based in large measure on the work of ecologist Robert M. May, a professor of zoology at Oxford University, Chairman of the Board of the Natural History Museum in London, and currently science advisor to Prime Minister Tony Blair. The estimate of thirty thousand species lost per year comes from the work of biologist E. O. Wilson, acknowledged dean of biodiversity issues, recently retired from Harvard University.

9. page 161
See my *Life in the Balance* (1998) for an extended examination of the present-day biodiversity crisis, including its causes, why we should care, and what we can do to stem the tide of this Sixth Extinction. And, when in New York, once again I invite you to visit the Hall of Biodiversity at the American Museum of Natural History—our first "issues" hall, devoted to the beauties of the living world, the importance of the living world to human life, the threats that it faces, and the solutions we might find to ending the crisis.

10. page 163
See Turnbull's "Cultural Loss Can Foreshadow Human Extinctions: The Influence of Modern Civilization" (1985).

11. page 164
In my book *Dominion* (1995), p. 100.

12. page 166
Earlier in this narrative, I mentioned debating Phillip Johnson at Calvin College in Grand Rapids, Michigan. The only instance when I can say I was challenged with some degree of animus was over my remark—also in my book *Dominion* (1995)—that it seems obvious to many of us that we created God in our own image, rather than the other way around. To some, this remark is offensive. Yet there is nothing intrinsic about that remark that says that the concept of God in question—the God of the Judeo-Christian tradition—does not exist in precisely the manner Christian theology specifies (however varied that characterization may be from theologian to theologian). It does imply, however, that in acknowledging that concepts of gods are ideas, mental images of such gods are precisely that: human constructs. Nothing wrong with that, and nothing about the statement, once again, states or implies that such gods either do or do not exist.

13. page 166
The only exceptions occur when species are reduced in size to a single population within a single ecosystem, as is most likely to occur in the very earliest and very latest phases of a species' existence.

# Bibliography

Behe, Michael J. 1996. *Darwin's Black Box: The Biochemical Challenge to Evolution.* New York: Free Press.

Bird, Wendell. 1978. *Acts and Facts.*

Bird, Wendell. 1991. *The Origins of Species Revisited: The Theories of Evolution and Abrupt Appearances,* 2 vols. Nashville, Tenn.: Thomas Nelson.

Brett, Carlton E., and Baird, Gordon. 1995. "Coordinated Stasis and Evolutionary Ecology of Silurian to Middle Devonian Faunas in the Appalachian Basin. In *Speciation in the Fossil Record,* R. Anstey and D. H. Erwin (eds.), pp. 285–315. New York: Columbia Unversity Press.

Bronowski, Jacob. 1973. *The Ascent of Man.* Boston: Little, Brown.

Buckland, William. 1823. *Reliquiae Diluvianae; Or, Observations on the Organic Remains Contained in Caves, Fissures, and Diluvial Gravel, and on Other Geological Phenomena, Attesting the Action of an Universal Deluge.* London: John Murray.

Crick, Francis. 1981. *Life Itself: Its Origin and Nature.* New York: Simon and Schuster.

Cuvier, Georges. 1817. *Essay on the Theory of the Earth.* Edinburgh: William Blackwood.

Dalrymple, Brent G. 1991. *The Age of the Earth.* Stanford, Calif.: Stanford University Press.

Darwin, Charles R. 1859. *On the Origin of Species by Means of Natural Selection.* London: John Murray.

Dawkins, Richard. 1976. *The Selfish Gene.* New York: Oxford University Press.

Dawkins, Richard. 1986. *The Blind Watchmaker: Why the Evidence of Evolution Reveals a Universe.* New York: W. W. Norton.

Dawkins, Richard. 1996. "Reply to Phillip Johnson." *Biology and Philosophy,* 11, 539–540.

Dobzhansky, Theodosius. 1937. *Genetics and the Origin of Species*. New York: Columbia University Press.

Doolittle, W. Ford. 1980. "Revolutionary Concepts in Cell Biology." *Trends in Biochemical Science*, 5, 146–149.

Dover, Gabriel, and Flavell, R. B. (eds.). 1982. *Genome Evolution*. London: Published for the Systematics Association by Academic Press.

Doyle, Arthur C. 1912. *The Lost World*. London: Hodder and Stoughton.

Easton, W.H. 1960. *Invertebrate Paleontology*. New York: Harper and Brothers.

Eldredge, Niles. 1971. "The Allopatric Model and Phylogeny in Paleozoic Invertebrates." *Evolution*, 25, 156–167.

Eldredge, Niles. 1981. "Creationism Isn't Science." *New Republic*, April 4, 15–17, 20.

Eldredge, Niles. 1982. *The Monkey Business: A Scientist Looks at Creationism*. New York: Washington Square Press.

Eldredge, Niles. 1989. *Fossils: The Evolution and Extinction of Species*. New York: A Peter N. Nevraumont Book, Harry N. Abrams.

Eldredge, Niles. 1995. *Dominion*. New York: Henry Holt.

Eldredge, Niles. 1998. *Life in the Balance: Humanity and the Biodiversity Crisis*. Princeton, N.J.: A Peter N. Nevraumont Book, Princeton University Press.

Eldredge, Niles. 1999. *The Pattern of Evolution*. New York: W. H. Freeman.

Eldredge, Niles. 2001. *Who Is This Man They Call God?* New York: A Peter N. Nevraumont Book, W. H. Freeman.

Eldredge, Niles. In press-a. "The Sloshing Bucket: How the Physical Realm Controls Evolution." In *Evolutionary Dynamics* (Proceedings of Santa Fe Institute Conference), J. Crutchfield and P. Schuster (eds.).

Eldredge, Niles. In press-b. "Biological and Material Cultural Evolution: Are There Any True Parallels?" *Perspectives in Ethology*, 13.

Eldredge, Niles, and Gould, Stephen. 1972. "Punctuated Equilibria: An Alternative of Phyletic Gradualism." In *Models in Paleobiolog*, T. J. Schopf (ed.), pp. 82–115. San Francisco: Freeman Cooper.

Eldredge, Niles, and Tattersall, Ian. 1976. "Fact, Theory and Fantasy in Human Paleontology." *American Scientist*, 65, 204–211.

Endler, John J.,. 1986. *Natural Selection in the Wild*. Princeton, N.J.: Princeton University Press.

Fisher, Ronald A. 1930. *The Genetical Theory of Natural Selection*. Oxford: Clarendon.

Freske, Stanley. 1981. *Creation/Evolution*.

Galton, Francis. 1867. *Hereditary Genius: An Inquiry into Its Laws and Consequences*. London: Macmillan.

Gingerich, Phillip. 1997. "The Origin and Evolution of Whales." LSA Magazine, University of Michigan, 20(2), 4–10.

Gish, Duane T. 1973. *Evolution: The Fossils Say No!* El Cajon, Calif.: Creation-Life Publishers.

Gosse, Philip. 1857. *Omphalos: An Attempt to Untie the Geological Knot*. London: John van Voorst.

Gould, Stephen J. 1977. *Ontogeny and Phylogeny*. Cambridge, Mass.: Harvard University Press.

Gould, Stephen J. 1989. *Wonderful Life: The Burgess Shale and the Nature of History*. New York: W. W. Norton.

Haeckel, Ernst. 1866. *Generelle Morphologie der Organismen*. Berlin: Georg Reimer.

Haldane, John B. S. 1931. *The Causes of Evolution*. New York: Longman, Green.

Hamilton, William D. 1963. "The Evolution of Altruistic Behavior." *American Naturalist*, 97, 31–33.

Hamilton, William D. 1964a. "The Genetical Evolution of Social Behavior, I. *Journal of Theoretical Biology*, 7, 1–16.

Hamilton, William D. 1964b. "The Genetical Evolution of Social Behavior, II. *Journal of Theoretical Biology*, 7, 17–32.

Hennig, Willi. 1966. *Phylogenetic Systematics*. Urbana: University of Illinois Press.

Hills, E.S. 1953. *Outlines of Structural Geology*, 3rd ed. London: Methuen.

Hopson, James A. 1987. "The Mammal-like Reptiles: A Study of Transitional Fossils." American Biology Teacher, 49, 16–26.

Horgan, John. 1996. *The End of Science: Facing the Limits of Knowledge in the Twilight of the Scientific Age*. Reading, Mass.: Addison-Wesley.

Hull, David. 1988. *Science as a Process: An Evolutionary Account of the Social and Conceptual Development of Science*. Chicago: University of Chicago Press.

Humphrey, Nicholas. 1992. *A History of the Mind: Evolution and the Birth of Consciousness*. New York: Simon and Schuster.

Hunter, George W. 1914. *A Civic Biology: Presented in Problems*. New York: American Book Co.

Hunter, George W. 1926. *New Civic Biology: Presented in Problems*. New York: American Book Co.

Hutton, James. 1795. *Theory of the Earth with Proofs and Illustrations*, 2 vols. Edinburgh, Scotland: William Creech.

Johnson, Phillip E. 1991. *Darwin on Trial*. Downers Grove, Ill.: InterVarsity Press.

Johnson, Phillip E. 1995. *Reason in the Balance: The Case against Naturalism in Science*. Downers Grove, Ill.: InterVarsity Press.

Johnson, Phillip E. 1996. "Response to Pennock." *Biology and Philosophy*, 11, 561–563.

Kevles, Daniel. 1985. *In the Name of Eugenics: Genetics and the Uses of Human Heredity*. New York: Knopf.

Kimura, Motoo. 1983. *The Neutral Theory of Molecular Evolution*. New York: Cambridge University Press.

Kirschvink, J. L., R. L. Ripperdan, and D. A. Evans. 1997. "Evidence for Large-Scale Reorganization of Early Cambrian Continental Masses by Inertial Interchange True Polar Wander." *Science*, 277, 541–545.

Lamarck, Jean-Bapiste 1809. *Philosophie Zoologique*. Paris: Dentu.

Larson, Edward J. 1997. *Summer of the Gods: The Scopes Trial and America's Continuing Debate Over Science and Religion*. New York: Basic Books.

Leuba, James H. 1916. *The Belief in God and Immortality*. Boston: Sherman French.

Lewin, Roger. 1982. "Judge's Ruling Hits Hard at Creationism." *Science*, 215, 381.

Linnaeus (Carl von Linné). 1758. *Systema Naturae per Regni Tria Naturae*. Stockholm, Sweden.

Lyell, Charles. 1830–1833. *Principles of Geology: Being an Attempt to Explain the Former Changes in the Earth's Surface by Reference to Causes Now in Operation*, 3 vols. London: John Murray.

Macbeth, Norman. 1971. *Darwin Retried: An Appeal to Reason*. Boston: Gambit.

Malthus, Thomas. 1798. *An Essay on the Principle of Population, As It Effects the Future Improvement of Society*. London: J. Johnson.

Margulis, Lynn (L. Sagan). 1967. "On the Origin of Mitosing Cells." *Journal of Theoretical Biology*, 14, 225–274.

May, Robert M., and Lawton, John H. (eds.). 1995. *Extinction Rates*. New York: Oxford University Press.

Mayr, Ernst. 1942. *Systematics and the Origin of Species*. New York: Columbia University Press.

McMenamin, Mark A. S. 1998. *The Garden of Edicara: Discovering the First Complex Life*. New York: Columbia University Press.

Mendel, Gregor. 1866. "Versuche über Pflanzen-Hybriden" (Experiments on Plant Hybrids). *Verhandlungen des naturforschenden Vereins, Abhandlungen, Brünn*, 4, 3–47.

Miller, Stanley M., and Orgel, Leslie E. 1974. *The Origins of Life on the Earth.* Englewood Cliffs, N.J.: Prentice-Hall.

Mivart, St. George. 1871. *Genesis of Species.* London: Macmillan.

Morris, Henry. 1963. *The Twilight of Evolution.* Grand Rapids, Mich.: Baker Book House.

Morris, Henry M. 1977. *The Scientific Case for Creation.* El Cajon, Calif.: Creation-Life Publishers.

Morris, Henry. 1980. "The Tenets of Creationism." *Acts and Facts,* 9(7).

Morris, Henry M., and Parker, Gary E. 1982. *What Is Creation Science?* El Cajon, Calif.: Creation-Life Publishers.

Morris, Henry M., and Whitcomb, John C. 1961. *The Genesis Flood: The Biblical Record and Scientific Implications.* Grand Rapids, Mich.: Baker Book House.

Morris, Simon C. 1998. *The Crucible of Creation: The Burgess Shale and the Rise of Animals.* New York: Oxford University Press.

Murchison, Roderick. 1839. *The Silurian System.* London: John Murray.

Neufeld, B. 1975. "Dinosaur Tracks and Giant Men." *Origins,* 2(2), 64–76.

Numbers, Ronald L. 1992. *The Creationists: The Evolution of Scientific Creationism.* New York: Alfred A. Knopf.

Osborn, Henry F. 1917. *The Origin and Evolution of Life: On the Theory of Action, Reaction, and Interaction of Energy.* New York: C. Scribner's Sons.

Palmer, Allison R. (ed.). 1982. *Perspectives in Regional Geological Synthesis: Planning for the Geology of North America.* Boulder, Colo.: Geological Society of America.

Parker, Gary E. 1980. *Creation: The Facts of Life.* El Cajon, Calif.: Creation-Life Publishers.

Pennock, Robert T. 1996a. "Naturalism, Creationism and the Meaning of Life: The Case of Phillip Johnson Revisited." *Creation/Evolution,* 16(2), 10–30.

Pennock, Robert T. 1999. *Tower of Babel: The Evidence against the New Creationism.* Cambridge, Mass.: MIT Press.

Pennock, Robert T. 1996b. "Naturalism, Evidence and Creationism: The Case of Phillip Johnson." *Biology and Philosophy,* 11, 543–559.

Raup, David M. 1991. *Extinction: Bad Genes or Bad Luck?* New York: W. W. Norton.

Read, J. G. 1979. *Fossils, Strata and Evolution.* Culver City, Calif.: Scientific-Technical Presentations.

Ross, C. P., and Rezak, Richard. 1959. *The Rocks and Fossils of Glacier National Monument.* U.S. Geological Survey Professional Paper 294-K.

Rudwick, Martin. 1997. *Georges Cuvier, Fossil Bones and Geological Catastrophes: New Translations and Interpretations of the Primary Texts.* Chicago: University of Chicago Press.

Ryan, William, and Pitman, Walter C. 1998. *Noah's Flood: The New Scientific Discoveries about the Event That Changed History.* New York: Simon and Schuster.

Schindewolf, Otto. 1993. *Basic Questions in Paleontology.* Chicago: University of Chicago Press.

Schuchert, Charles. 1955. *Atlas of Paleogeographic Maps of North America.* New York: John Wiley.

Simpson, George Gaylord. 1944. *Tempo and Mode in Evolution.* New York: Columbia University Press.

Simpson, George Gaylord. 1953. *The Major Features of Evolution.* New York: Columiba University Press.

Smith, William. 1816. *Strata Identified by Organized Fossils Containing Prints on Coloured Paper of the Most Characteristic Specimens in Each Stratum.* London.

Spencer, Herbert. 1866. *The Principles of Biology,* 2 vols. New York: D. Appleton.

Steno, Nicolaus (Niels Stensen). 1699. *De Solido intra Solidum Naturaliter Contento Dissertationis Prodromus.* Florence, Italy.

Sunderland, Luther. 1984. *Darwin's Enigma: Fossils and Other Problems.* El Cajon, California: Master Book division of CLP.

Sunderland, Luther. 1998. *Darwin's Enigma: Ebbing the Tide of Naturalism.* Santee, Calif.: Master Books.

Tattersall, Ian. 1998. *Becoming Human: Evolution and Human Uniqueness.* New York: Harcourt Brace.

Tattersall, Ian, and Schwartz, Jeffrey. 2000. *Extinct Humans.* Boulder, Colo.: A Peter N. Nevraumont Book, Westview Press.

Thompson, John J. 1994. *The Coevolutionary Process.* Cambridge, Mass.: Harvard University Press.

Turnbull, Colin. 1985. "Cultural Loss Can Foreshadow Human Extinctions: The Influence of Modern Civilization." *In Animal Extinctions: What Everyone Should Know,* R. J. Hoage, pp. 175–192. Washington, D.C.: Smithsonian Institution Press.

Vrba, Elisabeth. 1985. "Environment and Evolution: Alternative Causes of the Temporal Distribution of Evolutionary Events." *South African Journal of Science,* 8, 229–236.

Vries, Hugo de. 1910–1911. *The Mutation Theory: Experiments and Observations on the Origin of Species in the Vegetable Kingdom,* 2 vols. London: Kegan Paul.

Waagen, W. 1869. "Die Formenreihe des Ammonites subradiatus." *Beneckes Geognos t.-paläontol. Beitr.,* 2, 179–256.

Weber, Christopher. 1980. "Common Creationist Attacks on Geology." *Creation/Evolution*, 2.

Weidenreich, Franz. 1943. "The Skull of *Sinatropus pekinesis:* A Comparative Study of a Primitive Skull." (Palaeontologia Sinila New Series D, no. 10.) Pehpei, ChungKing: Geological Survey of China.

Weismann, August. 1893. *Germ Plasm: A Theory of Heredity.* London: Walter Scott.

Wertheim, Margaret. 1997. *Pythagoras' Trousers: God, Physics, and the Gender War.* New York: W. W. Norton.

Wertheim, Margaret. 1999. *The Pearly Gates of Cyberspace: A History of Space from Dante to the Internet.* New York: W. W. Norton.

Whewell, William. 1837. *History of the Inductive Sciences from the Earliest to the Present Time.* London: John W. Parker.

Willams, George C. 1966. *Adaptation and Natural Selection: A Critique of Some Current Evolutionary Thought.* Princeton, N.J.: Princeton University Press.

Wilson, Edward O. 1978. *Sociobiology.* Cambridge: Harvard University Press.

Wilson, Edward O. 1992. *The Diversity of Life.* Cambridge, Mass.: Harvard University Press.

Wright, Sewall. 1986. *Evolution: Selected Papers.* Chicago: University of Chicago Press.

Wysong, R. L. 1976. *The Creation-Evolution Controversy.* Midland, Mich.: Inquiry Press.

# Acknowledgments

Production of *The Triumph of Evolution* has been truly a team effort. I thank John Michel of W. H. Freeman—who had the idea in the first place and has been stimulating and encouraging throughout; Peter N. Nevraumont, Ann Perrini, Simone Nevraumont, and Ruth Servi Zimmerman (Nevraumont Publishing Company) for their kaleidoscopically varied editorial and production skills and support; Stephanie Hiebert, world-class copyeditor, who has come to know every word of this book!; and our design team Patrick Seymour and Kevin Smith of Tsang Seymour Design, who gave the book its look, and who came up with the best cover I've ever seen!

# Index